T0220285

Cambridge Elements ≡

Elements in the Philosophy of Mathematics
edited by
Penelope Rush
University of Tasmania
Stewart Shapiro
The Ohio State University

WITTGENSTEIN'S PHILOSOPHY OF MATHEMATICS

Juliet Floyd
Boston University

CAMBRIDGE
UNIVERSITY PRESS

CAMBRIDGE
UNIVERSITY PRESS

University Printing House, Cambridge CB2 8BS, United Kingdom

One Liberty Plaza, 20th Floor, New York, NY 10006, USA

477 Williamstown Road, Port Melbourne, VIC 3207, Australia

314–321, 3rd Floor, Plot 3, Splendor Forum, Jasola District Centre, New Delhi – 110025, India

103 Penang Road, #05–06/07, Visioncrest Commercial, Singapore 238467

Cambridge University Press is part of the University of Cambridge.

It furthers the University's mission by disseminating knowledge in the pursuit of education, learning, and research at the highest international levels of excellence.

www.cambridge.org
Information on this title: www.cambridge.org/9781108456302
DOI: 10.1017/9781108687126

First published 2021

A catalogue record for this publication is available from the British Library.

ISBN 978-1-108-45630-2 Paperback
ISSN 2399-2883 (online)
ISSN 2514-3808 (print)

Wittgenstein's Philosophy of Mathematics

Elements in the Philosophy of Mathematics

DOI: 10.1017/9781108687126
First published online: July 2021

Juliet Floyd
Boston University
Author for correspondence: Juliet Floyd, jfloyd@bu.edu

Abstract: For Wittgenstein, mathematics is a human activity characterizing ways of seeing conceptual possibilities and empirical situations, proof and logical methods central to its progress. Sentences exhibit differing "aspects," or dimensions of meaning, projecting mathematical "realities." Mathematics is an activity of constructing standpoints on equalities and differences of these. Wittgenstein's Later Philosophy of Mathematics (1934–51) grew from his Early (1912–21) and Middle (1929–33) philosophies, a dialectical path reconstructed here partly as a response to the limitative results of Gödel and Turing.

Keywords: Wittgenstein, rule-following, philosophy of mathematics, history and philosophy of mathematics, logicism, meaning in mathematics, surveyability, realism in mathematics, mathematical knowledge, logical necessity, mathematical necessity, language and mathematics, foundations of mathematics, philosophy of logic, logic and communication, mathematics and language, naturalism in philosophy of mathematics, philosophy of mathematical practice

JEL classifications: A12, B34, C56, D78, E90

ISBNs: 9781108456302 (PB), 9781108687126 (OC)
ISSNs: 2399-2883 (online), 2514-3808 (print)

Contents

1 Introduction

1.1 Aims and Sources

Few have characterized Wittgenstein's philosophy of mathematics over the entirety of his intellectual life. Some favor the Early and Middle Wittgenstein *over* the Later. Below I argue that Later Wittgenstein, a dialectical revisor of Early and Middle Wittgenstein, offers us a defensible contemporary philosophy of mathematics.

In 1944 Wittgenstein wrote that his "chief contribution has been in the philosophy of mathematics."[1] He aimed at a book whose first part would clarify the nature of meaning and whose second part would apply that clarification to the foundations of logic and mathematics. RFM is an edited selection from unfinished drafts Wittgenstein hoped would form PI part 2. But he withheld its publication. These writings utilize Wittgenstein's Later interlocutory style, but lack polished orchestration; interpreters have frequently distorted his ideas by using the method of drive-by quotation. Selected remarks lack their original context. The tentativeness of Wittgenstein's thoughts is obscured. Since 2000, BEE has been used to supplement RFM with manuscripts. We rely on commentaries using BEE.[2]

1.2 Aspect Realism

Wittgenstein sees mathematicians articulating conceptual constructions that provide standpoints for modeling empirical and mathematical facts, situations, structures, events, procedures, and characterizations. While its techniques are intertwined with logical features of language, mathematics is not reducible to a single "foundation" such as second-order logic or set theory. Later Wittgenstein suggests looking at mathematics as a "MULTICOLORED mix of techniques" (RFM III §§46,48) to nuance the kind of anti-foundationalism he always urged.

In Wittgenstein's Later Philosophy, aspects are *discovered* or revealed, whereas mathematical techniques are *invented*.[3] Mathematics is *both* discovery and invention. I read Wittgenstein as an "aspect realist" about mathematics.

This is a "realistic" form of realism in the sense of Diamond (1991) and Putnam (1999): realism without grounded metaphysics and no particular epistemology or theory of mind. It transposes what is often called "realism." I shall not worry whether the informal language of "seeing aspects" is metaphorical

---

[1] Monk, 1990, 466.
[2] Mühlhölzer, 2010 (RFM III); WH (RFM II, V).
[3] BT, §134; RFM II §38, RFM III §§46ff; MS 122, pp. 15, 88-88, 90; PI §§119,124-129,133, 222, 262, 387, and 536; xi, p. 196; PPF xi, §130. Floyd, 2018a; Harrington, Shaw, and Beaney, 2018; and Baz, 2020 concern aspects generally.

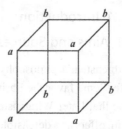

Figure 1 Necker Cube
Source: TLP 5.5423.

or wholly misjudged. The phenomenon is clear enough in the ambiguous depth cues of Jastrow's Duck-Rabbit and the Necker Cube (Figure 1).

Aspects are modal, attaching to possibilities and necessities: fields of significance, opportunities for projecting and instantiating our concepts. We see *through* the picture to our own seeing of it *as* realizing one way among others. What we see is seen, but also we see. We rearticulate what we see, sometimes seeing it thereby anew. There is an active and a passive aspect to this. Aspects show themselves (the middle voice).[4] What we are seeing is not simply an actual drawing on a page. We can also "see" *in* these drawings *possibilities* of projecting our concepts. Here we take modality as primitive, though up for investigation.

Wittgenstein always rejected the "Platonist" idea that mathematics reveals actual entities, abstract facts, or realms that explain our mathematical practices but are causally inert. Aspectual realism is an offered substitute. He has no story about whether a possibility is *itself* a possible state of affairs; he is burying ontology in that sense, as many current philosophers of mathematics do.[5]

Aspects shift the "naturalism" urged in psychologistic, conventionalist, and Humean readings of Later Wittgenstein: these underplay his Later views on truth, confusing them with assertability conditions.[6] Since Benacerraf (1965), many philosophers of mathematics have aimed to balance three main ideas: (a) an epistemological position privileging immediate perception of material objects, conceived as a causally cemented, receptive lining up of experience with words or concepts; (b) phenomena such as incompleteness, undecideability, and nonstandard models; (c) modality as something to be explained in metaphysical terms.

---

[4] Narboux, 2014.
[5] Auxier, Anderson, and Hahn, 2015, 11; Putnam, 2004.
[6] Fogelin, 1987, 2009; Maddy, 2014.

Wittgenstein rejects these ideas, while respecting their deep appeal. He broached a phenomenological account of mathematics in his early Middle Period as a way to resist them (§3.2). This failed, leading him to reject the idea that experience is intrinsically structured in a life-world. His Later Philosophy seeks to dislodge the argument between the naturalistic, causal view and the phenomenological one: both sides underrate the power of ordinary, colloquial ways of speaking in and about mathematics.

Though logicians have not failed to attend to the role of ordinary language (§3.1), the signature element in every period of Wittgenstein's philosophy is his reflection on our "silent adjustments" of its complexities (TLP 4.002). The Later philosophy finally factors these adjustments into realism itself.

Psychological contingencies are relevant to the question, "What is the nature of logic (and mathematics)?" Psychology explores aspect-perception as a phenomenon. But psychology cannot substitute for critical discussion of our ordinary discussion. When Wittgenstein remarks that a proof must be "surveyable," he is not discussing the width of our cognitive abilities but our need for *mathematics* (§4.4). His concerns about extensionalism (§3.5, §4.5) are in place even though, so far as I know, cognitive psychologists have no idea how to account for our coming to grasp the concept of *set*.

Realistically speaking, in mathematics there is moulting and molding of concepts: truth is not simply a matter of grasping an extension, asserting or stipulating a principle. Aspect "perceptions" show us "the limits of empiricism" in *Wittgenstein*'s sense.[7] This contrasts with typical mathematicians' images of these limits. Russell (1936) argued that the limits lie solely in the "medical" fact of our finitude and the need for abstraction to universals: we cannot drink an infinite number of glasses of water. But this is a contingent fact. Hrbacek and Jech (1999, 86), a textbook in set theory, states, for example, that

> it is possible to write down decimal expansions $0.a_1a_2a_3 \dots$ where "$\langle a_i \rangle_{i=1\dots\infty}$" is an arbitrary sequence of integers between 0 and 9.

Most readers pass over such remarks silently. Later Wittgenstein urges, not that such "supertasks" are incoherent, but that the limits of empiricism lie elsewhere: first, in our need to *communicate proofs* (RFM III §71), therefore, second, in the activity of *concept-formation* (RFM IV §29), and third, in our embeddings of words and symbols (including the above-quoted ones) in *forms of life* (RFM VII §§17, 21).

---

[7] RFM III §71, IV §29, VII §§17, 21.

Maddy (2011, 2007) advocates"thin" realism about sets and a Second rather than a First philosophy. "Aspect realism" is a form of this. But Maddy focuses first and foremost on set theory. Moreover, she lacks confidence in our ability to articulate a sense of depth in mathematics. Aspect realism is intended to round out "thin realism" to convey more of a 3D sense of mathematical depth and insight.

"Naturalistic" philosophers of mathematics have tended to privilege physics as the arbiter of ontology, and psychology as the basis for "naturalized epistemology."[8] In Wittgenstein a more "liberal" form of naturalism is in view, the kind advocated by the post-1990 Putnam.[9] Like Wittgenstein, Putnam held that mathematics explores conceptual possibilities (he called this "modal structuralism"). But he insisted on the normative elements in play here. "Forms of life" express the human animal's ways of structuring possible lives with language, and this has an evolutionary, biological tint, but also a normative and ethological one.[10]

The Later Wittgenstein emphasizes the importance of projectability and plasticity: the work of fitting concepts *to* reality, including the reality of mathematical (and other) experiences. This form of realism not only glosses the wide and multifarious kinds of applicability of mathematics – to empirical situations, to mathematics itself, to experiences, and to concepts. It also allows mathematics its autonomy, as modal structuralism does. Wittgenstein treats mathematics, however, more thinly: as a kind of *scaffolding* for descriptions, that is, a modular, transportable collection of possible conceptual constructions that may be configured and reconfigured in an unlimited variety of ways. It does not support the edifice of knowledge so much as help human beings erect it.

What we learn in mathematics comes to feel so natural, in certain cases, that it comes to shape our immediate experience, embedding its modalities deeply in our habits and perceptions. There is an echo of Kant here. Wittgenstein does not, however, forward a view of intuition as a fixed form of immediate, singular, nonconceptual representation. The idea instead is that there are mathematical *possibilities* and *necessities*, "forms" of particular, immediate experience, construction, and conceptualization typical of human beings and communicable among them.

Wittgenstein is *grammaticalizing* the "intuitive," that is, subjecting it to discussion, using the ordinary language of mathematics to do so, taking this language as (an evolving) given. He does not replace mathematical experience

---

[8]　Maddy, 1997, 2005.
[9]　Putnam, 1990, 1992, 2012, 2016.
[10]　Cavell, 1988.

with language, but rather uses language to open up our receptiveness *to* it. "Language itself provides the necessary intuition" (TLP 6.233). This allows him to capture "the reflective element" that Bernays (1959) missed in RFM. In the Later Philosophy, it is as if the whole idea of a metalanguage is absorbed into the ever-shuffling process of formalization, reformulation, renewed experience, informal characterization, re-parametrization, and reinterpretation. We should respect the hustle.

I detect an affinity with a Gödelian norm that plays a crucial role in mathematical practices, lately developed by Kennedy (2020). She calls it *formalism-freeness*. In Wittgenstein this is a matter of characterizing what formalisms are and mean, ideally with a minimal degree of formalization. Both Gödel and Wittgenstein associate aspect-richness with incompleteness. Whereas Gödel hypothesizes that there may be an actual infinity of complexity given in experience, Wittgenstein regards the unbounded degree of complexity as potential and grammaticalizable.[11] He stresses the *non*-extensional rather than the extensional perspective. While Gödel felt the perspectives could be conceptually merged, Wittgenstein resisted this.

I shall use the term "non-extensional" instead of the more usual term "intensional," which is typically associated with constructivism as a working branch of mathematics. I do not regard Wittgenstein as a constructivist about mathematics, and he did not believe in intensions as entities.

Aspect-talk allowed Wittgenstein to soberly discuss the *experience* of novelty, the reorientation of our way of seeing that comes when we encounter mathematical features, forms, and structures anew.[12] There is a two-ness of grammatical complexity in the entanglement of modality with the ways we use the verb to "see" and "seeing-as" phrases to describe it: through characterizations we can come to "see" or "reveal" or "discover" a possible way of conceiving of something we hadn't seen before. Mathematics is filled with such "seeing." This is not literal perception, but more like seeing a new dimension or possibility for thought.

Wittgenstein takes our as- and aspect-phrasing to serve us essentially in our ability to draw meaningful distinctions between characterizations and properties, discoveries and inventions, appearances and realities, possibilities and necessities. Aspects open up space for the *kind* of discoveries it is possible to make in philosophy, logic, and mathematics – and also the kinds of things that people may *miss*. That there is such a thing as insight into things and people, into possibilities that are not necessities, into truths and falsehoods, and

---

[11] Floyd and Kanamori, 2016, MS 163, 40v.
[12] RFM I App II (Mühlhölzer, 2002) criticize unsober ideas of discovery.

that such things are not merely fleeting appearances or conventions – these are among Wittgenstein's deepest philosophical themes.

Aspect is therefore a logical notion. It reworks Frege and Russell. For them, truth is absolute; there is no *way* of being true.[13] For Wittgenstein there are different ways of articulating truth *as* an absolute notion.

The term "aspect" does not occur in Frege's German, though English translators have often used it to render his remarks on the decomposition of judgments. We hear Frege speak (1879, §9) of our "imagining that an expression can be altered" at a place in a sentence; in shifting "$2^4 = 16$" to "$x^4 = 16$" the term is "regarded as replaceable" (Bauer-Mengelberg) and we may "imagine the 2 in the content" so that the content is "split into a constant and a variable part" (1880/1881, 17–19), though Frege actually is careful to say that it divides, or decomposes itself. In

$$(1 + 1) + (1 + 1) + (1 + 1) = 6$$

"the different expressions correspond to different conceptions [aspects] and sides, but nevertheless always to the same thing" (1891, 4f.); a name "illuminates its reference only one-sidedly" (1892, 27).

These imaginings, regardings, and 3D metaphors flirt with procedural and with psychologistic language, even though Frege is marking something static: the generality of a concept. Wittgenstein's aspect-talk picks up on this. The slide is between a metaphor of proceeding, "thinking" a position *as* variable, and a metaphor of something exposed (a concept or function). Frege himself warned against this metaphorical danger (1897, 157n), (1918/19, 66).

In TLP Wittgenstein emphasizes the multiple "standpoints" (aspects as prospects, or views) from which the sameness or difference of meaning of a sentence or arithmetic expression might be assessed (6.2323). Mathematics consists of equations and uses the "method of substitution" (6.234ff.). Logical patterns – shared "internal" features of sentences – evince "faces" or "looks" (4.1221). But Wittgenstein replaces Frege's idea of "varying a position" in a sentence by the idea of possible step-by-step "operations": formal procedures that evince "internal" necessity. He operationalizes generality and meaning.

Like Frege, Wittgenstein resists the idea that the process of decomposing a thought into its components requires "intuition." But he rejects Frege's general sense/reference distinction. For Wittgenstein (Early, Middle and Later), names have *Bedeutung* (reference), but not sense; only sentences, in which names structurally figure, have sense (*Sinn*), evince aspects, help project the realization of specific possible situations.

---

[13] Shieh, 2019.

Wittgenstein's aspect-phrasings invite the charge that he confused talk about essence with essence, sign with object, truth with appearance. Russell's use of aspects in his (1914) construction of the world had urged something close to this. But Wittgenstein was pursuing the reverse idea, committed to fulfilling the analytical role played by Russell's notion of "acquaintance": immediate, direct knowledge of a mind-independent, particular thought-form. In responding to Russell, Wittgenstein returns the concept of "acquaintance" to its everyday home, the sense in which one may be "acquainted" with a person, a proof, a face, a habitat: one looks, speaks, walks around, thinks, from different angles. This de-emphasizes the causal story by recasting the role of "perception."

For Wittgenstein, mathematics may be said to be "about" numbers, aspects of concepts, and so on, but only in an ordinary language sense familiar from Austin.[14] This emphatically does *not* mean that "all we are talking about when we do mathematics is language" or that Wittgenstein confused linguistic and metaphysical matters. Ordinary language is about whatever we talk about: so it is not *just* the words, it is how we look at things, describe things, *fit* our concepts *to* one another and reality. What is "ordinary" is not simply given; rather, it is what is familiar, taken for granted. It must be thought *through* – occasionally denied or shifted – for its potentialities (aspects) to be seen.

A number, as Gowers nicely puts it, "*is* what it *does*" (2002, 18). But it does not do what it does, have the force and reality and necessities and possibilities it does, without being fit for (indefinite) elaboration in specific ways – like any action, human or animal. An action is no action at all unless it falls under a series of intelligible characterizations. An action, a number, a proof have different *aspects*, that is, necessities, faces, characters.[15]

These phenomena are real. We can regard the real number field *as* a field; a proof *as* a model of an experiment (RFM I); we may see (or miss) a grasp of the concept "counting" *in* a child's behavior. This projection and reassembling of concepts is, in fact, ubiquitous.

2 Early Philosophy (1912–1928): Absolute Simplicity

2.1 "Final" Analysis

TLP urges an ideal of *absolute* simplicity: the design of a logical notation in which the totality of sentences ("what can be said") is formally constructed, step-by-step, from a base of simplest "elementary" sentences, atomic configurations of "simple" names that "picture" facts. Possibilities of combinations

---

[14] Mühlhölzer, 2014.
[15] Diamond, 1991; Floyd, 2010, 2012a; Putnam, 2012.

of names in elementary sentences would mirror the possible configurations of facts. Wittgenstein uses schematic language – "$f(x)$","$\phi(x,y)$", "p", and so on (4.24) – but this draws out necessary "formal" features of any language, not specific values in a particular language from the point of view of a metalanguage. There is just *one* language that I understand, just *one* logical space.

Russell broached the possibility of a hierarchy of languages in his Introduction to the TLP. But Wittgenstein never accepted that reasoning in a metalanguage about a syntactically specified formal language clarified fundamental philosophical issues. If devoid of attachment to a working, meaningful language, a formal language is logically accidental. Wittgenstein distinguishes between *signs* (sign types, e.g., "A is the same sign as A" [3.203]) and *symbols* that are essential (logical) features of the expression of thought in any language.

The main task in rewriting ordinary sentences in a Tractarian notation is to separate what is arbitrary from what is not arbitrary in thought. Symbols receive a "formal" expression in the final analysis: it is unthinkable that an object or thought lacks an "internal" feature that its symbols so display (4.1221).

The TLP excludes a "theory" of types: the possibilities of saying what is the case inscribe types directly into the grammatical workings of the symbols. What it is impossible to think is nonsense, and Wittgenstein has no theory of what that is. Russell's "paradox" is obviated by seeing that a concept such as $F(fx)$ cannot take itself as argument, because if this is done, the working symbolic "prototypes" automatically disambiguate: "$F(F(fx))$" should be rendered $\psi(\phi(fx))$ (3.333).

Wittgenstein's conception of second-order generalization is unclear. Do types automatically circumscribe the range of sentences demarcated by $\psi(\phi(fx))$? Did he take such a form to be constructible, step-by-step, from the forms of elementary sentences? On one reading we imagine free use of second-order generalization, as in Frege, with types carrying the load.[16] Alternatively only predicatively defined, recursive second-order concepts are allowed in the generation of second-order variables.[17]

An intermediate position envisions a type-free language with a λ-operator and a cumulative, predicative presentation setting out the language in stages: second-order variables shift their range depending upon the (finite) stage of the construction of the language.[18] This expresses a portion of *Principia*'s ramified type theory: eventually every second-order "form" appears, its instances restricted from below.

---

[16] Potter, 2000; but compare Weiss, 2017, 7.
[17] Ricketts, 2014; Sundholm, 1992.
[18] Fisher and McCarty, 2016.

Possibilities of logic are not states of affairs that we picture, but come through – are "shown" in – our saying what is the case and our thinking, truly or falsely. We have the capacity to regard our sayings as affirming and/or denying the realization of this or that set of possibilities among others, leaving certain possibilities aside as logically independent. Logic explores the formal interrelations among (structured) sayings of what is the case, rewriting "ordinary" sentences in a notation to reveal "necessary" relations amongst them, even those involving "material" concepts (e.g., empirical laws).

In the imagined final analysis, each elementary sentence, being logically independent of all the others (as in truth-functional logic), would be capable of being affirmed or denied, rightly or wrongly (each is true or false). A particular affirmation of "everything that is the case" would affirm/deny each elementary sentence in a manner consistent with the totality of its consequences, accomplishing in action a truth-functional assignment of T's and F's to each elementary sentence. Wittgenstein's point is that same-saying by affirmation and/or denial is definite and unambiguous.

Wittgenstein's Operator N yields a step-by-step, formal construction of all possible sentences. The totality of the results of N's constructive applications unify the formal aspects of logic and language in what Wittgenstein calls the "general form of sentence" (hereafter GFS). The TLP's notational devices are all devoted to making this rewriting project seem not only plausible but necessary.

Sentences are composed of symbols exhibiting dimensions of generalization on which we can formally "operate" in a step-by-step manner: aspects. If I think, for example, that the object a is red (suppose a and "red" appear in the final analysis), then a logical aspect of the thought is expressed with the sentential variable "x is red." A general form, "$\phi(x)$," ranges over all elementary sentences sharing this form.

"Simple objects" are the ultimate nodes of inference in the final analysis. The forms of their possible configurations mirror all possible "internal" – that is, logical – relations among structures of thought, the common "form" and "substance" of all ways of thinking or imagining what might be so. An "internal" relation among thoughts p and q (or objects) is necessary in that they cannot fail to have and remain the thoughts (objects) they are: that is, p is logically equivalent to q. Crystalizing the totality of both form and content, undefinable by further analysis or description, the names of simples are like coordinates and dimensions constituting "logical space." Philosophy (logic) is to construct "a system of signs of a definite number of dimensions – of a definite mathematical multiplicity" (5.475).

The final analysis would fully disambiguate language: different names indi-
cate different objects. Thus the identity sign is eliminated from the proposed
notation as redundant (5.53f.), with the cardinality of the universe expressed
by the number of names – obviating the need for an *Axiom* of Infinity. Russell's second-order definition of identity as indiscernibility is unnecessary
(5.5302ff.), his theory of descriptions reduced to configurations of names.
Frege's treatment of sense and reference for names vanishes (6.232).

The right "mathematical multiplicity" would emerge a posteriori, as a result
of research (5.55ff.). It is not part of the task of analysis to anticipate the actual
compositions of sentential structures needed to represent reality, for example, the number of n-place relations or the extent of higher-order analysis or
geometry required in physics.

Yet Wittgenstein urges that a complete a priori analysis of all the pure possibilities and mathematical multiplicities *must* be possible. We are capable of
setting aside any factual truths, laws, or happenstances by seeing *through* our
everyday activities of picturing facts *to* what is the case *if* a thought or sentence is true. Sentences are themselves facts, structures, none of which is a
priori true, each of which may be compared with the facts. Form is the *possibility* of structure, the being-capable-of-being-true-or-false, or "sense" (*Sinn*)
of a sentence (2.033).

This transposes the traditional dialectic between Realism and Idealism
("Skepticism") into a purely logical key. Analysis presents the truth, the whole
truth, and nothing but the truth, *simply* – that is, determinately and clearly (3.23,
5.4541, 5.5563). It is coherent to imagine that all our affirmations of what is
the case are false, that they are all true, or that some are true, some are false.
Logic and mathematics run through all possibilities of thought. This is a move
reminiscent of Kant, except that while Wittgenstein remarks that logic is "transcendental," he also holds the un-Kantian view that all logic is general (formal)
logic, which is "analytic" and therefore "tautological."[19]

Like Kant, Wittgenstein never regarded mathematical knowledge as tautological. Aspects are his logicist-inspired way of articulating the "syntheticity"
of thought at which Kant aimed. The Vienna Circle did not follow the philosophy of mathematics of the TLP, for they assumed that the logicist reduction
of mathematics to logic was successful. Carnap, who attempted to reconstruct
arithmetic and analysis as tautological and failed, relabeled the truths of logic
and mathematics "analytic" but *not* "tautological," reserving the latter term, as
we now do, for the smaller class of truth-functional tautologies.

---

[19] Dreben and Floyd, 1991.

2.2 The General Form of Sentence, Form Series

The GFS expresses a rule for constructing all sentences using iterated applications of N, a generalized form of a Sheffer Stroke: not only single sentences but *collections* of sentences and/or their forms may be jointly negated by its means.

Wittgenstein mentions three ways generality may appear through N's application (5.501). First, N may be applied to a finite list of sentences to produce a list of their negations. Second, "$N(f(\overline{x}))$)" expresses the collection of "all" sentences which are negations of sentences of the form of a propositional variable such as "$f(x)$" (the bar indicates "all"). This gets us to quantificational generality. Wittgenstein interprets his translation of Russellian quantifiers in such a way that different variables bound by the same quantifier must be instantiated by different values, in accordance with the complete disambiguation of names in the final analysis (4.1272). This makes it difficult to translate his notation into ours. After a decades-long dispute, Rogers and Wehmeier (2012) settled that Operator N can express first-order logic with identity. However, Wittgenstein's logic is more powerful than this. For it also contains, third, the generality of recursive definitions.

"Form series" collect *series* of sentence forms by means of terms expressing the potentially infinite application of an iterated, step-by-step, formal rule (5.501):

> The general term of a form series a, $O'a$, $O''a$, …. I write thus: "$[a, x, O'x]$".
> This expression in brackets is a variable. The first term of the expression is the beginning of the form series, the second the form of an arbitrary term x of the series, and the third the form of that term of the series which immediately follows x.

A "formal law" does not represent a material concept but an *operation* characterizing relations among forms: a well-founded formal field of variation on a single rule. Unlike a sentence or sentential function used in the expression of a material concept (e.g., "$f(x) \lor g(x)$"), a form series variable does not belong directly to the logical form of any particular sentence: it is elliptical, a nonextensional device. Wittgenstein sharply differentiates its generality from the merely "accidental" generality of a set or class (6.031).

Any particular enumeration of the elementary sentences of a language will exhibit "accidental" generality: the elementary sentences are logically independent, so any starting point is arbitrary. But what is not arbitrary is that *some* enumeration of all sentences is possible. It might be that the number of elementary sentences, each logically independent of all the others, is (countably) infinite. But each of them is a finite structure (a determinate picture, a fact).

No matter what the size of the set of elementary sentences (finite or infinite), an arbitrary correlation of syntax with a (potentially infinite) alphabet must be possible.

Why? Wittgenstein *defines* numbers as "exponents" of formal operations enumerating the number of steps taken in the construction of a formal series, no matter what the base of the series and what the operation (6.02, §2.4). This possibility of indexing includes subscripting a (lexicographically ordered) alphabet of names to structure a formally ordered collection $(a_1, a_2, a_3, \ldots)$. Thus may the elementary sentences be enumerated. Given any psychological, auditory, or pictorially expressed thoughts, it will be possible to enumerate them using translations into an alphabetical language (4.014, 4.025). (Gödel later showed the possibility of arithmetically coding the well-formed formulas of any formalized countable language.)

Any enumeration of the elementary sentences is arbitrary, as it utilizes "accidental features" of the language involved (3.34): particular names go proxy for objects according to different conventions in different languages. But in any language something is not arbitrary. It is the *way* the names are configured in a particular language that constitutes their expression of a thought, their picturing of a specific fact: sentences are not mere collections of names (3.1432). And thus what is not arbitrary is the *possibility* of enumerating the sentences beginning with *some* articulation of the language.

What is presupposed in this claim are the capacities to (a) begin from a particular base (of simple names and their combinatorial possibilities in elementary sentences), (b) distinguish identity and difference of names, and (c) construct a formal series based on the identity of outcome of applying a formal rule to any reoccurrence of a particular name or sentential form. Let us call (b) and (c) the "Uniformity Principle." For Poincaré (1904, 1.I) it is logical, not mathematical:

> It must be granted that mathematical reasoning has of itself a kind of creative virtue, and is therefore to be distinguished from the syllogism. The difference must be profound. We shall not, for instance, find the key to the mystery in the frequent use of the rule by which the same uniform operation applied to two equal numbers will give identical results. All these modes of reasoning, whether or not reducible to the syllogism, properly so called, retain the analytical character, and *ipso facto*, lose their power.

Wittgenstein *logicizes* the Uniformity Principle, drawing it into the foundations of mathematics. What is not accidental to logic is the capacity to use form series terms to express thoughts in the form of the generality of formal laws *and* to index them. By "capacities" we do not mean present, actual abilities we possess, materially speaking, but rather "logical capacities." These are primitive.

Speaking extensionally, the first and second forms of generality (5.501) – (a) listing sentences and (b) using sentential variables for "material" concepts – are encompassed by (c) the generality of a form series term. However, for Wittgenstein the nonextensionalist there is a sharp contrast between (a) and (b) versus (c), the latter characterizing rules ordering forms. Presumably Wittgenstein held that every formal "law" may be expressed by a form series term. But not every collection of logical forms will be a form series.

The GFS is given in terms of a form series expression. N is the heart of the TLP: its expressive capacity is adequate to express all the truth functions, and it may be used to characterize all (possible) logical forms. Wittgenstein's idea is that the operation of negation runs through the formal unity of logic, all saying what is the case. Indeed, the "fundamental thought" of the TLP is that the logical constants do not name but characterize operational potentialities with expressions (4.0312). Bivalence and the law of excluded middle are not postulates or laws but reflected in the use of any representing language.

As for the GFS:

> The general form of truth-function is: $[\bar{p}, \bar{\xi}, N(\bar{\xi})]$.
> This is the general form of sentence.
> This says nothing else than that every sentence is the result of successive applications of the operation $N'(\bar{\xi})$ to the elementary sentences (6).

"$\bar{p}$" "stands for all collections of elementary sentences, "$\bar{\xi}$" for an arbitrary result of applying operator N a finite number of times to any such collection, and $N(\bar{\xi})$ for all possible applications of operator N to all possible results of such collecting and joint negating with N. "$N'(\bar{\xi})$" characterizes *the* result of applying N a finite number of times, step-by-step, to any such base. Wittgenstein's claim is that every affirmable sentence, under analysis, is such a result.

The form series term is Wittgenstein's most novel device. It gives his logic the expressive power of (at least) some predicative fragment of second-order logic. It allows him (at least) to define the ancestral of a relation, logical consequence, and all arithmetically definable sets.[20] More broadly, it draws into his philosophy the idea that it is part of our very *concept* of logic that we are able to write down – that is, express with a finite number of symbols – formal rules that we can follow step-by-step, thereby constructing a potential infinity of forms.

---

[20] Weiss, 2017.

2.3 Early Philosophy of Logic

Poincaré (1904) begins:

> The very possibility of mathematical science seems an insoluble contradic-
> tion. If this science is only deductive in appearance, from whence is derived
> that perfect rigour which is challenged by none? If, on the contrary, all the
> sentences which it enunciates may be derived in order by the rules of formal
> logic, how is it that mathematics is not reduced to a gigantic tautology?

Wittgenstein reworks Poincaré's "contradiction." Logic consists wholly of tautologies (and their negations, contradictions) (6.1201f.). A "purely logical"
conclusion is already "contained" in the premises of any deductive proof,
and every deductive proof is tautological when conditionalized. Logic carries zero information (6.1203), admitting no comparison with reality (4.461).
Logical "propositions" are without sense (*sinnlos*), though not nonsensical
(*unsinnig*).

Language has the power to cancel out its own procedures through grammar.
This is a signature theme for Wittgenstein. The TLP remarks that "the whole of
philosophy" is full of this: philosophical "propositions" are nonsense (3.324)
– including the remarks of the TLP, which are instead "formal" elucidations
(6.54,7). Axiomatization in logic is epistemologically superfluous. The number
of "primitive propositions of logic" in any system is arbitrary (6.1271), a matter
of accidental presentation: the GFS encompasses all we require. Proof *in* logic
is, in quantifier-free cases, a "mechanical" expedient for determining logical
validity (6.1262).

Axiomatization in mathematics is more than window dressing, however,
for we use it to titrate and compare different theories. Hilbert wrote of the
"half-timbering" of concepts involved in this enterprise: universally applicable
struts that hold the edifice of mathematical concepts together.[21] Wittgenstein's
alternative picture is that logic and mathematics involve *scaffolding*, formal
rearrangements (3.42). They are not conceptual frameworks, but rather modular, transportable, humanly used pieces of language that interconnect in a
variety of ways. They may be erected and transformed as we build the edifice of science, but they do not support that edifice. Wittgenstein replaces the
axioms of *Principia* with formal procedures. Axiomatizations run the danger of leading us to confuse tautologies or equational rules of arithmetic for
propositions depicting "reality." This may drive us to respond with conventionalism about logical aspects of mathematics. Wittgenstein aims to avoid these
dangers.

---

[21] Hilbert to Frege, December 29, 1899, in Frege, 1983, p. 60.

The sign "=" is extralogical, expressing the intersubstitutivity of operational forms. The method of mathematics is "the method of substitution" (6.24), a "logical method." This method is nonextensional, for it conveys the idea of *necessary* (rule-governed, procedural) generality. Wittgenstein took formal series – the potential infinity of step-by-step, operational routines, calculations, and substitution steps – as basic and ubiquitous elements of logic and mathematics. The extensional point of view, the theory of sets or classes, abstracts away from all procedures and rules, offering only "accidental" generality, the happenstance of a collection without order, something of no ultimate logical significance.

Frege and Russell were also nonextensionalists (Frege spoke of "concepts" and "senses," Russell of "propositional functions"). But they took the two perspectives, extensional and nonextensional, to fit together. By contrast, Wittgenstein drew a sharp distinction between the two. *Principia*'s Axiom of Reducibility states that for every nonpredicatively specified sentential function there exists a predicatively specifiable extensionally equivalent function of a lower order. For Wittgenstein this is logically inessential to mathematics, for it might turn out to be false (6.1232). Unlike the equations of elementary arithmetic, we must await the final analysis's verdict on whether or not it applies to forms of "reality."

To reason about infinite sets in mathematics, we *must* develop an extensional theory, abstracting away from all possible human procedures and rules. We appeal to rules, choices, and procedures also in set theory, but the extensionalist regards this as "merely" human. Wittgenstein works, by contrast, with an everyday notion of mathematical understanding and separates the perspectives. If one imagines another – supertasks or hypercomputability – this would be a matter of metaphysical accident rather than logical necessity.[22]

2.4 Early Philosophy of Mathematics

Like Poincaré, the TLP takes mathematical insight to be independent of the particular quantificational logical structures of Frege and Russell *because* logic consists of tautologies (6.233). Unlike Poincaré, Wittgenstein ties mathematical forms of operation to logical form as the *possibility* of structure. Wittgenstein proceduralizes the intuitive, substituting atemporal procedures for what Kant regarded, in his philosophy of arithmetic, as the formal intuition of succession in time. This expresses a debt to logicism and a rebellion against it.

---

[22] Russell, 1936; Copeland, 2002.

Wittgenstein's notion of *operation* is equiprimordial with the GFS:

> If we are given the general form of the way in which a sentence is constructed, then thereby we are also given the general form of the way in which by an operation out of one sentence another can be created.
> The general form of the operation $\Omega'(\bar{\eta})$ is therefore: $[\bar{\xi}, N(\bar{\xi})'](\bar{\eta})$ $(= [\bar{\eta}, \bar{\xi}, N(\bar{\xi})])$.
> This is the most general form of transition from one sentence to another.

No operation can generate a form that fails to be potentially attached to forms of elementary sentences. But there may be operations that bear an indirect relation to the representing power of language (e.g., substitutions). Mathematics consists of a "calculus" of operations spelling out possible forms of depicting reality. Its operations carry propositions to propositions, not primarily *via* the application of truth functions and/or quantificational rules, but more generally by means of iterable operations that are applicable to sentences, forms, and operations themselves.

Form series earmark Wittgenstein's lifelong resistance to the idea of a *reduction* of arithmetic to logic. He takes mathematics to provide us with ways of seeing forms of possible structurings *of* objects and procedures. His philosophy of mathematics is an ontologically neutral, predicative modal structuralism taking recursive procedures as basic, abstracting away from any particular factual basis. In mathematics there are no empirical hypotheses. Yet the applicability of mathematics in the logical forms of empirical science is indispensable.

No matter what the mathematical multiplicity in the final analysis, it is possible to write down, formally, *any* number using form series (6.431). If the number of objects turns out to be finite, then the operational language of arithmetic reflects a certain redundancy, like clock arithmetic.

The arithmetical language Wittgenstein envisions in the TLP is, on the surface (though not in its potential applications), "logic-free." Numbers are *indices* of steps taken in form series (6.2323), so their "generality" is not quantificational but given through the possibility of parametrizing formal procedures:

> 6.02 And thus we come to numbers. I define
>
> $$x = \Omega^{0'}x \text{ Def. and}$$
> $$\Omega'\Omega^{\nu'}x = \Omega^{\nu+1'}x \text{ Def.}$$
>
> According, then to these symbolic rules we write the series
>
> $$x, \Omega', \ \Omega'\Omega'x, \ \Omega'\Omega'\Omega'x, \ldots$$
>
> as:
>
> $$\Omega^{0'}x, \ \Omega^{0+1'}x, \ \Omega^{0+1+1'}x, \ \Omega^{0+1+1+1'}x, \ldots$$

Therefore I write in place of "$[x, \xi, \Omega'\xi]$",

$$"[\Omega^0{}'x, \Omega^{\nu}{}'x, \Omega^{\nu+1}{}'x]".$$

And I define:

$$0 + 1 = 1 \text{ Def.}$$
$$0 + 1 + 1 = 2 \text{ Def.}$$
$$0 + 1 + 1 + 1 = 3 \text{ Def.}$$
and so on.

A number is the exponent of an operation.

Frascolla (1997) reconstructs Wittgenstein's language rigorously, writing down the "proof" of $2 \times 2 = 4$ that Wittgenstein gives (6.241) and noting the similarity of the formalized language to that of the λ-calculus.[23] However, in that calculus "λ" works as a function. Wittgenstein's "Ω's" are schematic, indicating possible parametrizations of formal aspects of language, mathematical and extramathematical.

In labeling arithmetic equations "pseudosentences" Wittgenstein is protecting its universal *applicability* (i.e., its readiness to accommodate any set of elementary propositional forms). He accounts for the "formal" character of, for example, "if I have 2 eggs and 4 eggs, then I have 6 eggs" as substitutional, thereby satisfying the logicist demand that inferences using ascriptions of number be nonempirical. However, Wittgenstein takes such inferences to be analogous to – though not identical with – tautologies. In construing equality as the expression of a formal and procedural relation he rejects the idea that an analysis of ascriptions of number ("there are 6 eggs") involves reference to numbers as objects or an identity claim concerning properties of concepts treated extensionally (*via* the notion of a one-to-one correlation).

How does Wittgenstein provide a nonquantificational account of our knowledge of *general* truths of number and arithmetic, including the validity of proofs by mathematical induction? These seem, as Poincaré insisted, not to be anything tautology-like. Inductive arguments contain within them a potentially infinite cascade of steps of reasoning from the holding of a property $\phi(x)$ for each particular number n to $\phi(Sx)$ for its successor Sn (Poincaré, 1904, 10). This rule seems to presuppose, in turn, that we may infer $\phi(n)$ from $(\forall x)\phi(x)$.

Wittgenstein treats the potentially infinite number of conclusions as a "synthetic" form of reasoning that is formal (5.43). What Frege–Russell logic renders in terms of universal instantiation is aspectual in Wittgenstein. Induction is conditionalized reasoning insofar as particular numbers await the final analysis,

---

[23] Compare Marion, 1998 and Fisher and McCarty, 2016.

but it is direct as a formal possibility. In any inductive proof all the mathematical work goes into establishing that there is a formal "projection" from the possession of property $\phi(x)$ by n into "the same" for Sn.

This is why mathematical induction is not even mentioned in the TLP: Wittgenstein takes its mathematical "validity" to emerge directly from the form of any form series. The number series 0,1,2,3,... is rendered as a formal series term "$[0, \xi, \xi + 1]$" in which "ξ" ranges over exponents of operations (6.022, 6.03). Its "generality" is procedural (aspectual), depending upon mathematical insight and the three presuppositions of form series generality mentioned above in our gloss of 5.501 (§2.2).

Wittgenstein's emphasis on the gap between extensionalism and nonextensionalism complicates his inheritance of Russell and Frege. They too insisted that symbolic logic reflects nonextensional aspects of meaning (e.g., "concepts" and "propositional functions"), but their axioms crossed the boundaries between the two viewpoints (Frege's Basic Law V; Russell's Axiom of Reducibility). Wittgenstein refuses this. He always resisted construing equinumerosity extensionally, as the actual *existence* of a one-to-one correlation between the objects falling under them.[24] For him cardinality and equinumerosity are aspectual, evincing the possibility of "projecting" one aspect of collection into another *through* the possibility of indexing elements with formal series.

In *Principia* the Axiom of Reducibility is used to define identity second-order and to develop real analysis: it guarantees that impredicative definitions (those using quantifiers ranging over the classes they characterize) pick out sets belonging to same type. Since Wittgenstein does not regard identity as a genuine relation, the first task is otiose. Dedekind's analysis of real numbers as "cuts," or classes of rational numbers, uses impredicative definitions to define basic concepts such as the "least upper bound" of an interval of reals that is bounded from above. This requires Reducibility to guarantee the membership of that number in a set characterized at a lower type. Wittgenstein allows for Reducibility to be used as a principle in mathematics for certain conceptual purposes internal to the extensionalist perspective. But this evinces "accidental" generality and is mathematically "superfluous" (6.02). In "real life," he remarks, we "*only*" utilize mathematical propositions to help us with material inferences (6.211).

There is no characterization of real numbers or the continuum in the TLP. Like Frege, Wittgenstein regarded Dedekind's cuts as unable to account for the applicability of real numbers.[25] The TLP does not contain much discussion of

---

[24] Sullivan, 1995; Floyd, 2005, 98f.
[25] Frege, 1893/1903, II §83; Snyder and Shapiro, 2019.

geometry – only an insistence that, like arithmetic, its forms are potentially varied enough to handle any of the particular logical forms that may arise through analysis: "a place is a possibility: something can exist in it" (3.032f., 3.411). Geometry evinces all-pervading, necessary formal features of our portrayals of reality. But logical "space" is shaped by it in a specific way only in the final analysis, where it would be determined, for example, whether material space is or is not Euclidean.

In TLP the arithmetic of numbers (whole, integer, and presumably real numbers) as well as the measurement relations of geometry are built into the very idea of *any possible* language. But Dedekind's "cuts" and set theory are not. In fairness to Wittgenstein, the presentation of real analysis in textbooks through the 1920s was nonextensional: a matter of developing calculational techniques for functions of a real variable.[26] Physics's uses of analysis depend neither on the Dedekindian analysis of what a real number *is* nor on set theory. The Axiom of Choice's status was not yet crystallized in mathematical practice at the time.[27] Dedekind himself did not regard his "cuts" as an analysis of the continuum but rather a possibility projected *into* the number line through an axiom that is incapable of proof.

2.5 Poincaré's Objections

The Uniformity Principle is a procedural, nonextensional counterpart of the notion of *function* (it speaks of "inputs" and "outputs"). The TLP shows that the "frequent use" of this principle is more fertile, leads to more novel insight, than Poincaré believed. The identity of results of rules developing formal aspects of language through operations is ubiquitous in mathematics, as well as logic. Uses of our number terms as parameters in constructing contexts for proof and calculation everywhere rely upon its applications.[28]

Wittgenstein explored the status of the Uniformity Principle throughout his life; what evolved was his understanding of the nature of its fundamentality. In modalizing it, he did not endorse an *empirical* axiom of the stability of signs.[29] But his departure from Frege's and Russell's logicism necessitates a reply to two main objections.

First, for Frege and Russell any particular symbolic articulations used in the presentation of deductive logic are not essential to the content of the justifications of arithmetical knowledge and logical relationships that logic provides.[30]

---

[26] Hobson, 1927, WH §§7ff., §4.3 below.
[27] Waismann, 1936a, chapter 13; WH Chapter 8.
[28] Shapiro, 2008, 2012 discusses ontological, semantic, and pragmatic aspects of parametrization.
[29] Compare Frege, 1884, xxf.
[30] Goldfarb, 2018b.

One may count parentheses and use mathematical induction over signs and formulas to read the symbolism, but this is irrelevant to the mathematical knowledge that the language of logic articulates – as irrelevant as is the color of the ink involved. Our uses of the formal signs involves a psychological under-standing of how we cognitively process logic, but this lies outside Frege's and Russell's purview as logicists. Frege made a "reproof" of Wittgenstein dur-ing their early conversations, viz., that "Wittgenstein places too much weight upon signs."[31] Wittgenstein himself worries about "the danger" of his get-ting "entangled" in "inessential" psychological investigations with his remarks about notation (4.1121).

The TLP riposte is that any particular notation *as* a notation does not matter to logic or mathematics. What matters is what is common to, what is reflected in, all *possible* notations (3.3421). Logic deals with all *possible* modes of thought, not the laws of its actual empirical realizations. The fact that a particular scheme for calculating with tautologies *can* be devised shows something logical, just as the possibility of a particular scheme for enumerating the sentences of a language shows something logical (§2.2). It is not the sign-manipulations qua causal events that matter, nor the psychological capacity we have to reidentify shapes visually. It is rather a question of the *possibilities* of our activities with the signs. This *itself*, and not any particular logical principles, is at stake. Witt-genstein thus sidesteps the logicist accusation that he has confused signs with justifications.

But what about the law of excluded middle and hence the *consistency* of all the possible applications of logical inference in Frege's and Russell's systems? For them there is a gap here, and perhaps a more sophisticated version of the Poincaré objection applies.[32] For they did not present logic in our contemporary way, in terms of a syntax presented from the standpoint of a metalanguage, or in terms of an axiomatic system having models.[33] Rather, Frege called his *Begriffsschrift* a "formula *language*" (1879).

Given the universality of this language (and of *Principia*, for Russell), nei-ther Frege nor Russell had the means to prove that a particular sentence does *not* follow by means of logic from other sentences: their systems can only dis-play positive results. Wittgenstein aimed in TLP to develop a general notion of "follows by logic" (logical consequence) that would verify or falsify *any* claim

---

[31] Floyd, 2011, 7.

[32] Parsons, 1983, 170f.; I disagree with Parsons and Waismann, 1936a, chap. 9 that Later Wittgenstein regards "$F(0)\&(\forall n)(Fn \supset F(Sn))$" as a "*criterion*" for the truth of "for all natu-ral numbers n, Fn" that *assures* us of no conflict between the two methods for proving individual cases; that is *Middle* Wittgenstein's view.

[33] Van Heijenoort, 1967.

about validity or consequence (just as truth tables do). (Later on Gödel [1930] provided a proof procedure for first-order validity while Church [1936] and Turing [1936] showed there is no decision procedure for this.)

The issue of consistency only became pressing for Wittgenstein as he developed a response to Hilbert's writings on the foundations of mathematics in the 1920s. Early Wittgenstein simply assumes a well-founded, step-by-step, recursive construction of sentences. Being mutually logically independent from one another, the elementary sentences are assumed to form a consistent set and are bivalent (if they appear not to be so, analysis is not complete).

The second issue distancing Wittgenstein's logicization of Poincaré from mere psychologism is the sharp distinction he draws between experiment and calculation as human epistemological processes: something Poincaré (1904, Preface) also emphasizes. In an experiment, we do not know the outcome ahead of time, we await the results and test hypotheses (4.031, 5.154). In a calculation, by contrast, we regard the result as nonaccidental, timelessly contained in the premises: "process and result are equivalent" and "there are no surprises," not in a psychological sense, but in a logical sense of necessity (6.1261, 6.2331).

Consider probability (5.1ff.).[34] For any (fair) pair of dice we are capable of "extracting" from a list of the total possible outcome-configurations probabilistic "features" of their likely behavior (5.154f.). We can, for example, calculate the probability of throwing a particular sum of the dice by dividing the total number of ways to throw that total by the total number of possible combinations it is possible to roll (viz., 36). We calculate, for example, that the probability of throwing a sum of 5 on a particular throw is 11.11 percent. Probability works with "formal" features of the total possibility space and, like arithmetic and geometry, serves as a *standard* against which empirical events are measured.

Wittgenstein sharply distinguishes such calculations from the application of probability in everyday life: "What can be confirmed by experiment, in sentences about probability, cannot possibly be mathematics" (NB p. 27, 8.11.14). What we "know" in probability statements are "certain general [formal] properties of ungeneralized sentences of natural science" – ultimately these turn on the elementary sentences. The general laws of physics attach to them in the final analysis. But this is not yet to know what will happen in everyday life on a particular roll of the dice.

Wittgenstein likens such applications of mathematics to our experiences of aspects of puzzle pictures. We may spot a tiny rabbit in the corner of a complicated picture where at first it was not visible. Or we might "see" the solution of

---

[34] Compare Floyd, 2010.

how to put puzzle pieces together to form a particular shape. The mathematical structure may "click" into place with what we experience, the "reality" of the formal concepts appear to be exemplified. This is not something empirical, and it involves something other than calculation. It exercises a capacity we have for looking at the world, modeling or seeing a happening or experience or sequence of events *in a particular way*, under an *aspect* of (mathematical) necessity.

TLP likens this to spotting an expression (say, an emotion) in a picture-face (4.1221). The emotion, like the logical feature, is internal to the particular structure of the face: when we see a smile in a happy face we see the apparent manifestation of *happiness*, one amongst other emotions in a field of various possible ones. When we apply aspect-perception to someone's face in everyday life we may mistake one emotion for another, just as when we apply probability, we may not see things "click" (NB 15.10.16). But the happiness is still a "formal" feature of the face.

Here emerges a crucial analogy for Wittgenstein, one he retains for the rest of his life. It is not highlighted in TLP, but powerfully incipient (4.1221) and grows in importance over time. We are capable of "seeing the face of necessity" *in* a human procedure or empirical process, seeing the "reality" (i.e., the possibility of structure) of a concept.[35] This is what makes the applications of logic and mathematics possible. "Seeing" the necessity does not mean ascribing a material property to objects, or imposing a convention upon our descriptions, or seeing a further fact or object behind the empirical events. It is instead a matter of fitting the necessities of a concept *to* empirical events. We are able to see *through* processes and situations, however falteringly, to see *in* them necessities and possibilities. This is a power, not of direct intuition of forms or necessary truths, but of reflection on aspects of our own experiences and encounters with the world. Philosophy is the development of this power.

2.6 Interpreting Form Series

For Frege and Russell, the notions of "all concepts" and "all objects" were reflected in the second-order quantificational notation of logic; for Wittgenstein, "material" concepts reflect empirical science, and logic and mathematics construct calculative routines capable of accommodating any set of material concepts science might provide. The elementary sentences are *directly* elementary: they exhibit no "hierarchy" of types; rather, any "hierarchies" are constructed by us via operator *N*. Wittgenstein remarks that we can only "foresee" what we ourselves construct (5.556): he regards logic and mathematics

---

[35] Diamond, 1991, 243ff.

as constructible in that we can "foresee" that their constructions involve form, not content, and so run through all possible thought about what is the case. This leaves ample room for the "creative" aspects of mathematics that differentiate its forms from those of logic, for we cannot foresee which forms of calculative manoeuvres and parametrizations it may be necessary to provide for experience.

Wittgenstein agrees with Weyl (1918, 48) that *"the idea of iteration, i.e., of the sequence of the natural numbers, is an ultimate foundation of mathematical thought."* But he interprets "sequence" non-extensionally, in terms of the possibility of a procedure. This prevents him from defining the ancestral of a relation in the Dedekind–Frege–Russell, top-down, second-order way. Like Poincaré and Weyl, he charges that this technique for changing an elliptical procedure into an explicit definition contains a *"circulus vitiosus."*[36] He offers instead a form series (4.1273):

> If we want to express in logical symbolism the general sentence "*b* is a successor of *a*" we need for this an expression for the general term of the form series:
>
> $$aRb,$$
> $$(\exists x) : aRx.xRb,$$
> $$(\exists x, y) : aRx.xRy.yRb,$$
> $$\cdots \cdot$$

Weiss (2017), relying on what he calls a "minimalist" reading of the use of form series, shows how one may use Wittgenstein's form series to frame all finitary inductive definitions, the ancestral, and all arithmetically definable relations, as well as categorical finite axiomatizations of arithmetic. Weiss notes (p. 44) that the complexity of the consequence relation for this logic is sensitive to the size of the domain; on his reading, the precise formulation of the consequence relation would be settled for Wittgenstein through analysis, that is, only in the application of logic.

This does not resolve the question of what, in general, Wittgenstein's form series representations allow. Weiss's form series use operations allowing substitutions of sentences for sentences and sentential forms, beginning from a sentence as base. Converting a name in an elementary sentence into a variable (e.g., "*Aa*" to "*Ax*", "*Ax*" to "*Aa*") counts, as does substituting a sentence into the place where another sentence is in a more complex schema. A "minimalist" form series (MFS) results from replacing each occurrence of sentence *p* in

---

[36] In Whitehead and Russell, 1910, second edition, Russell attempted to characterize "natural number" predicatively, but this cannot be done: Marion, 1998, 43.

complex schema B with some base sentence or first-order sentential variable and then, in subsequent steps, again with a sentence or (first-order) sentential variable A. Weiss represents this as

$$[\![A, \xi, B[p/\xi]]\!].^{37}$$

As an example, consider the following form series spelling out

$$[\![Aa, \xi, [Aa \wedge /\xi]]\!],$$

namely,

$$Aa, Aa \wedge Aa, Aa \wedge (Aa \wedge Aa), \ldots .$$

We use parentheses to mark procedural groupings implicit in Wittgenstein's idea of a form series term (the associativity of operation results is a "formal" feature of the intersubstitutivity of expressions (6.23)). Associativity in arithmetic requires no axiom or proof by induction; it is built into the fact that number terms are signs for possible indexings of formal series and already contain their (procedural) generality of application.[38]

There must be something more than "minimal" operations that are plausibly "formal" in Wittgenstein's sense. Working with a base sentential variable expression, consider the spelling out of two further MFS's, $[\![Ax, \xi, [/\xi \wedge Ax]]\!]$:

$$Ax, (Ax) \wedge Ax, (Ax \wedge Ax) \wedge Ax, \ldots$$

and $[\![\neg Ax, \xi, [/\xi \wedge Ax]]\!]$:

$$\neg Ax, \neg Ax \wedge (\neg Ax), \neg Ax \wedge ((\neg Ax) \wedge \neg Ax), \ldots .$$

Since numbers are exponents of operations, we can index the variables produced at each step of the formal series:

$$Ax_1, Ax_1 \wedge (Ax_2), Ax_1 \wedge ((Ax_2) \wedge Ax_3)), \ldots$$
$$\neg Ax_1, \neg Ax_1 \wedge (\neg Ax_2), \neg Ax_1 \wedge ((\neg Ax_2) \wedge (\neg Ax_3)), \ldots$$

A Tractarian justification for the multiple occurrences of (interior) variables in these series is that identity and difference of signs has to be primitive for identity-free signs, and the Uniformity Principle must apply to the results of applying operations to particular forms taken as base. (Because of this, as

---

[37] Weiss, 2017, 16.

[38] This answers the objection Frege makes to Leibniz's attempt to show that the truths of arithmetic may be proven from definitions (1884, §6); compare Potter, 2000, §2.1.

$$\Omega^{1,} x, \ \Omega^{2,} x, \ \Big| \ \Omega^{3,} x, \ \Omega^{4,} x,$$

$$\Omega^{1,} x, \ \Omega^{2,} x, \ \Big| \ \Omega^{3,} x, \ \Omega^{4,} x \ \Omega^{5,} x.$$

Figure 2 Odd and even

$$(\Omega^{1,} x, \ \Omega^{2,} x, \ \Omega^{3,} x, \ \Omega^{4,} x \ \Omega^{5,} x)$$
$$- (\Omega^{1,} x, \ \Omega^{2,} x)$$
$$= \qquad (\Omega^{1,} x, \ \Omega^{2,} x, \ \Omega^{3,} x).$$

Figure 3 Subtraction

we have already argued above, an enumeration of the alphabet of the language must be possible.) One also presumes that each application of the formal law involves a substitution of a variable different from the variables already occurring in preceding formulas of the series (in the TLP, differing variables must be instantiated with different values if they occur in "the same" context).

Every number-form is involved in the *formal* possibility of indexing repeated applications of operations for "constructing" form series indefinitely many times ("all" is understood here non-extensionally, not quantificationally), regardless of what the cardinality of the actual universe turns out to be.

Further mathematical concepts may be developed. "x is even" and "x is odd" emerge from seeing whether an ordered series can be equally divided in half (Figure 2). Subtraction? Lay two form series beside one another and re-index (Figure 3). An account of the integers emerges from adding a mirror image, reflecting. Given the idea of a *possible* procedure that is fixed in logical space by the simples, there is a kind of "geometry" of operation signs available to us.

Such a "geometry," using our indexing and re-indexing of form series, implicitly appeals to an operation that "dovetails" multiple form series expressions. This requires something different from MFS's, namely the construction of a new formal series from the combining of two others.

For example, consider the formation of a formal series term expressing "There are n A's," for all natural numbers n.

The initial term of such a series is a formula such as $A(x)$, with x a variable free in A. Choosing for simplicity's sake an enumeration of those variables

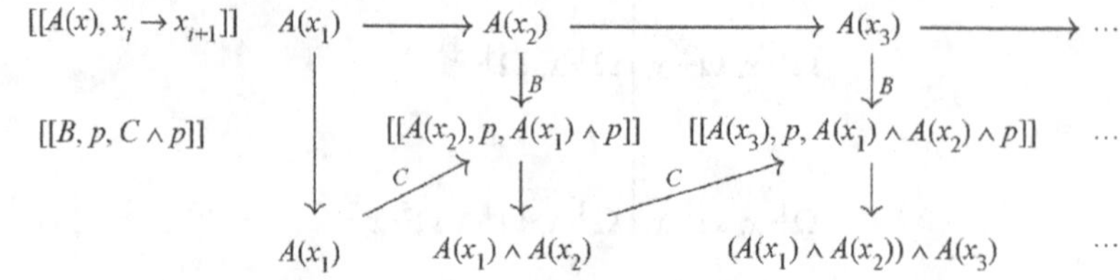

$[[A(x), x_i \to x_{i+1}]]$ $A(x_1) \longrightarrow A(x_2) \longrightarrow A(x_3) \longrightarrow \cdots$

$[[B, p, C \wedge p]]$ $[[A(x_2), p, A(x_1) \wedge p]]$ $[[A(x_3), p, A(x_1) \wedge A(x_2) \wedge p]]$ $\cdots$

$A(x_1)$ $A(x_1) \wedge A(x_2)$ $(A(x_1) \wedge A(x_2)) \wedge A(x_3)$ $\cdots$

Figure 4 Dovetailing formal series

$x_i, i > 1$ that do not occur in A, and setting x_1 to be x, the operation or formal law is to substitute for at least one occurrence of the free variable x_{i-1} in A the next variable x_i:

$$A(x_1), A(x_2), A(x_3), \ldots$$

and for this we could use the following notation:

$$[\![A(x), x_i \to x_{i+1}]\!].$$

Next, we dovetail with the "schematic" minimalist form series $[\![B, p, p \vee C]\!]$. The process constructs a series of form series; we then calculate first-step outputs of each member of this series of series to produce a new form series (Figure 4).

The base steps of each term in the series of form series terms are the results of substituting the terms of the first series, in order, for each basis. The formal rule expressions in the series of form series emerge by substituting for "C" the previous result of applying "$C \wedge p$", then moving on to the next basis for the next form series. One "calculates" one value and then proceeds to the next substitution into a basis. The whole process constructs the new form series:

$$A(x_1), A(x_1) \wedge A(x_2), (A(x_1) \wedge A(x_2)) \wedge A(x_3), \ldots.$$

A further form series may then be constructed by dovetailing with a series of generalized applications of Operator N to each term in this form series, one at the first step, two at the second, and so on:

$$N_{x_1}(A\overline{x_1}), N_{x_2}(N_{x_1}(A(\overline{x_1}) \wedge A(\overline{x_2}))), N_{x_3}(N_{x_2}(N_{x_1}((A(\overline{x_1}) \wedge A(\overline{x_2})) \wedge A(\overline{x_3})))), \ldots.$$

This expresses a formal series of denials of ascriptions of numbers $1, 2, 3, \ldots n$ to sentential function Ax. It may be translated into more familiar notation using our quantifiers and Wittgenstein's convention of excluding instantiation by the same name of different variables within the scope of a single quantifier:

$$\neg(\exists x_1)(Ax_1), \neg(\exists x_1, x_2)(Ax_1 \wedge Ax_2), \neg(\exists x_1, x_2, x_3)(Ax_1 \wedge Ax_2 \wedge Ax_3), \ldots. \quad (1)$$

$$\mathbf{a}_{11} \quad \mathbf{a}_{11}a_{12} \quad \mathbf{a}_{11}a_{12}a_{13} \quad \mathbf{a}_{11}a_{12}a_{13}a_{14} \quad \mathbf{a}_{11}a_{12}a_{13}a_{14}a_{15} \cdots$$
$$a_{21}\mathbf{a}_{22} \quad a_{21}\mathbf{a}_{22}a_{23} \quad a_{21}\mathbf{a}_{22}a_{23}a_{24} \quad a_{21}\mathbf{a}_{22}a_{23}a_{24}a_{25}$$
$$a_{31}a_{32}\mathbf{a}_{33} \quad a_{31}a_{32}\mathbf{a}_{33}a_{34} \quad a_{31}a_{32}\mathbf{a}_{33}a_{34}a_{35}$$
$$a_{41}a_{42}a_{43}\mathbf{a}_{44} \quad a_{41}a_{42}a_{43}\mathbf{a}_{44}a_{45}$$
$$a_{51}a_{52}a_{53}a_{54}\mathbf{a}_{55}$$

Figure 5 Diagonalization

We can now apply generalized Operator N to negate each term in this form series, generating a series of ascriptions of "at least" $1, 2, 3, \ldots n$ to the sentential function Ax:

$$(\exists x_1)(Ax_1), (\exists x_1, x_2)(Ax_1 \wedge Ax_2), (\exists x_1, x_2, x_3)(Ax_1 \wedge Ax_2 \wedge Ax_3), \ldots . \quad (2)$$

How to refine this to "exactly n"? We do this by combining the denials with the affirmations of ascriptions of number according to the following diagonal routine – a use of formal series construction of which Wittgenstein was undoubtedly aware.

First, represent the terms of the series (1) and (2) by writing them in an array, placing the form series terms of (2) on the top row. According to the needed schematic indexing, "a_{11}" is the first term of (2), "a_{12}" its second term, "a_{21}" is the first term of (1), and so on:

$$a_{11}a_{12}a_{13}a_{14}a_{15} \cdots$$
$$a_{21}a_{22}a_{23}a_{24}a_{25} \cdots$$
$$\vdots$$

Next, create a new form series by conjoining the diagonal terms in each rectangle stage $j_1, \ldots j_n$. Conjunction is attained by disjoining the negations of the n diagonal terms $a_{1j} \ldots a_{jj}$ and applying Operator N to jointly deny the result (Figure 5).

This gives us the formal series for ascriptions of exactly the number n, for all natural numbers $1, 2, \ldots, n$, to Ax, expressing Frege's adjectival strategy for ascriptions of number in terms of a form-series variable:[39]

$$(\exists x_1)(Ax_1) \wedge \neg(\exists x_1, x_2)(Ax_1 \wedge Ax_2) \ldots ,$$
$$(\exists x_1, x_2)(Ax_1 \wedge Ax_2) \wedge \neg(\exists x_1, x_2, x_3)(Ax_1 \wedge Ax_2 \wedge Ax_3), \ldots ,$$
$$(\exists x_1, x_2, x_3)(Ax_1 \wedge Ax_2 \wedge Ax_3) \wedge \neg(\exists x_1, x_2, x_3, x_4)(Ax_1 \wedge Ax_2 \wedge Ax_3 \wedge Ax_4), \ldots$$

---

[39] Frege, 1884, §56; Potter, 2000, §2.5.

$$\mathbf{a_{11}} \quad a_{11}\mathbf{a_{12}} \quad a_{11}\mathbf{a_{12}}a_{13} \quad a_{11}a_{12}a_{13}\mathbf{a_{14}} \quad a_{11}a_{12}a_{13}a_{14}\mathbf{a_{15}} \cdots$$
$$\mathbf{a_{21}}a_{22} \quad a_{21}\mathbf{a_{22}}a_{23} \quad a_{21}a_{22}\mathbf{a_{23}}a_{24} \quad a_{21}a_{22}a_{23}\mathbf{a_{24}}a_{25}$$
$$\mathbf{a_{31}}a_{32}a_{33} \quad a_{31}\mathbf{a_{32}}a_{33}a_{34} \quad a_{31}a_{32}\mathbf{a_{33}}a_{34}a_{35}$$
$$\mathbf{a_{41}}a_{42}a_{43}a_{44} \quad a_{41}\mathbf{a_{42}}a_{43}a_{44}a_{45}$$
$$\mathbf{a_{51}}a_{52}a_{53}a_{54}a_{55}$$

Figure 6 Ascriptions of number

If we wish to include 0, we index differently, making (1) the top row, so that now "a_{11}" is the first term of (1) "a_{12}" its second term, "a_{21}" the first term of (2), and so on. Then the series of ascriptions of number to a concept Ax, for all whole numbers $0, 1, 2, \ldots$,

$$\neg (\exists x)(Ax),$$

$$(\exists x_1)(Ax_1) \wedge \neg(\exists x_1, x_2)(Ax_1 \wedge Ax_2),$$

$$(\exists x_1, x_2) \wedge \neg(\exists)(x_1, x_2, x_3)(Ax_1 \wedge Ax_2 \wedge Ax_3) \ldots,$$

may be represented by conjoining terms in stages (Figure 6).

We may then index *these* stages of procedure and recombine the indexed variables into one unindexed variable "x" to neaten the presentation, using numbers as indexes of the final stages of operation:

$$(\exists_0)Ax, (\exists_1)Ax, (\exists_2)Ax, (\exists_3)Ax, \ldots.$$

These freewheeling uses of formal series and "applications" of number terms to index form series expand the repertoire of MFS's considerably. One sees how heavily one depends upon the Uniformity Principle as one moves from stage to stage in each "diagonal" construction.

We were inspired to these constructions by a puzzling pair of pictures Later Wittgenstein wrote in RFM II §11 (see §4.3 below):

Since my drawing [of the diagonal schema at RFM II §1] is only an *indication* of infinity, why must it be like this

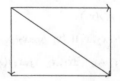

and not like this

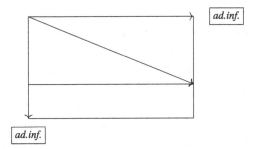

Here what we have is different pictures; and to them correspond different idioms. But does anything useful emerge if we have a dispute about the justification of *them*? What is important must reside elsewhere; even though these pictures fire our imagination most strongly.

The image is that there are alternative ways of knotting the a_{ik}-carpet of form series (Figure 7).[40] Now our speaking of "diagonals" no longer makes sense, but it is still the elements of the schematic form a_{ii} that are referred to when we construct the form series.

We may imagine countless alternatives to this process, with its successive augmentation of squares. These may well lead to "the same" a_{ik}-carpet and "the same" formal series. The important point for Wittgenstein is that a supposed actual infinity of the a_{ik}-array, conceived extensionally, may be replaced by potential infinity, understood *via* the processes mentioned, *together* with the idea of an equivalence of each with regard to their results *regarded as part of a procedure*. In this way, actual infinity, a notion belonging to the extensionalist

$$
\begin{array}{llllll}
\mathbf{a_{11}} & \mathbf{a_{11}}a_{12} & \mathbf{a_{11}}a_{12}a_{13} & \mathbf{a_{11}}a_{12}a_{13}a_{14} & \mathbf{a_{11}}a_{12}a_{13}a_{14}a_{15} & \cdots \\
a_{21} & a_{21}\mathbf{a_{22}} & a_{21}\mathbf{a_{22}}a_{23} & a_{21}\mathbf{a_{22}}a_{23}a_{24} & a_{21}\mathbf{a_{22}}a_{23}a_{24}a_{25} & \\
a_{31} & a_{31}a_{32} & a_{31}a_{32}\mathbf{a_{33}} & a_{31}a_{32}\mathbf{a_{33}}a_{34} & a_{31}a_{32}\mathbf{a_{33}}a_{34}a_{35} & \\
 & a_{41}a_{42} & a_{41}a_{42}a_{43} & a_{41}a_{42}a_{43}\mathbf{a_{44}} & a_{41}a_{42}a_{43}\mathbf{a_{44}}a_{45} & \\
 & & a_{51}a_{52}a_{53} & a_{51}a_{52}a_{53}a_{54} & a_{51}a_{52}a_{53}a_{54}\mathbf{a_{55}} & \\
 & & & a_{61}a_{62}a_{63}a_{64} & a_{61}a_{62}a_{63}a_{64}a_{65} & \\
 & & & & a_{71}a_{72}a_{73}a_{74}a_{75} & \\
\end{array}
$$

Figure 7 Alternative diagonalization
Source: WH Chapter 7.

---

[40] WH,179f.

point of view, obtains a non-extensional simulation. The constructive nature of diagonal method is drawn out, shifting our focus to the procedural *aspects* of the array: the rules as followed, rather than the extensions (see §4.3 below).

Poincaré expressed incredulity at the idea that the Uniformity Principle could yield an account of the creative and richly varied aspects in mathematics. Yet our brief foray into form series shows us that they give us the power to refashion new form series with seemingly endless shifts of newness.

As Ramsey (1923) pointed out, the TLP treatment of elementary arithmetic is minimal. It is not clear at all that or how Wittgenstein could formulate negation (inequations) or existence claims ("there are infinitely many primes"). If we mis-substitute one numeral and miscalculate, Wittgenstein's idea is that this will show itself in possible calculations elsewhere, resulting in an obvious violation of equality for a single numeral (WVC, 193–194). An inequation cannot be denied, as it has no sense. And what is the status of theorems such as "adding two odd numbers always yields an even number"?

This drives home the stark contrast between quantificational logical structure and mathematical structure in the TLP. Above all, Wittgenstein forwards a picture of the logical structure of ordinary language and how it *must* be if samenesses of outcome in step-by-step procedures are to hold. That is an assumption he will jettison.

3 Middle Philosophy (1929–1933): Relative Simplicity

Middle Wittgenstein engaged with Brouwer, Skolem, Weyl, Hilbert, and the Vienna Circle, especially Waismann. Conversations with Ramsey, beginning in 1923, intensified with his return to Cambridge in 1929. He received a fellowship (1930–6), lecturing and writing intensively.

Brouwer's (1929) lecture in Vienna apparently inspired Wittgenstein to return to philosophy, but he never subscribed to intuitionism.[41] While he sympathized with the radicalism of Brouwer, he endorsed no particular foundational approach. Ramsey sought to save classical mathematics from what he called the "Bolshevik" menace of Brouwer and Weyl, but to Wittgenstein this made him "a bourgeois thinker": Wittgenstein wanted to canvas "the foundations of any *possible* state," not only the then-dominant classical one.[42]

So Wittgenstein surrendered the idea of one ultimate notation with *the* correct mathematical multiplicity, relativizing simplicity to systems of sentences (*Satzsysteme*). These are mutually independent of one another analogous to the elementary sentences in TLP. They are "arbitrary," that is, not justifiable

---

[41] Menger, 1994; Marion, 2003, LFM, 237.
[42] MS,112, 70v; PI §125, Floyd, 2016, 44f.

a priori: standards of justification are internal to each system. Mathematics is now regarded as comprised of a plurality of discrete, autonomous "calculi" of sentences whose grammar fully determines their logical place in the system.

Aspects accommodate the idea that logical "features" are not merely *Züge* in the sense of formal truth-functional operations that "pull out" series of formal features by operations (4.1221), but also "grammatical" features of concepts involving multifarious uses of language. Wittgenstein still aimed to resolve philosophical conundrums, but soon abandoned the TLP's formal methods, expanding his repertoire to include a "genetic" method:[43] philosophical "contradictions" should be picked apart to see what patterns of thought lead *to* them – like psychoanalysis (PG 381-2).

Wittgenstein discussed Ramsey's (1926) and (1930) with the Vienna Circle (WVC, 129,189). The latter dealt with a portion of Hilbert's *Entscheidungsproblem*, which asked whether there was a general, "definite" method by which one might determine, in a finite number of steps, whether or not a particular sentence in a formalized theory does or does not follow from a set of axioms (Wittgenstein posed a version of this to Russell in 1913).[44] He rejected Ramsey's construal of this as a "leading" problem of mathematical logic (WVC 129, PI §124). The grammar (logic) of each *Satzsystem* internally and completely determines the sense, Yes or No, of every sentence in the system. Wittgenstein was denying that one "logic" could cover all *Satzsysteme* and hence that Hilbert's problem could be meaningfully posed.

Ramsey (1926) followed TLP, construing the theorems of *Principia* as tautologies, but advocated an extensionalist conception of identity and "propositional functions."[45] Wittgenstein resisted this, but took the difficulties of non-extensionalism to heart.[46] If mathematics depends upon actually infinite objects, then the extensionalist point of view is mandatory; yet it is difficult to see how to construe this dependence.

The TLP was extensionalist with respect to the construction of sentences using material concepts: Operator N included possibly infinite conjunctions and disjunctions. But Wittgenstein now rejected this, opting for non-extensionalism by rejecting the assimilation of quantifiers to truth-functions. The independence of the elementary propositions in TLP had allowed for the dispensability of identity through disambiguation of names, but now the complete

---

[43] Engelmann, 2013.
[44] Dreben and Floyd, 1991, 33.
[45] Potter, 2005.
[46] Sullivan, 1995.

disambiguation was carried out by relativizing names to their roles in *Satzsysteme*. Equations and other sentences of mathematics now have "sense," may be true or false, within a particular system (names do not have sense, their roles being exhausted by their combinatorial possibilities).

At Schlick's request, Wittgenstein began a collaboration with Waismann to produce an accessible presentation of the TLP (1929–30); but he resisted publishing a "rehash of [TLP] theses" (WVC, 183f.). The project was never completed. By 1932 he and Waismann turned toward coauthoring a new book; this cooperation failed with the rapid evolution of Wittgenstein's ideas. Waismann quickly produced accessible and insightful writings developing Middle Wittgenstein's views; these have retained an influence, particularly the idea of "open textured" concepts, a variant of Wittgenstein's Middle/Later idea of "family resemblance" (§4.1).[47]

3.1 Colloquial Language and "Effectiveness"

Effective and *computable* – step-by-step procedural reckoning by human beings working according to explicit rules – were unanalyzed, informal notions through the early 1930s. Yet the centrality of effectiveness, in practice, was central to Wittgenstein's treatment of logic and mathematics. Thanks to Turing's analysis of the idea of taking a "step" in a formal system (1936), the notion has been rigorized.[48] Turing's analysis succeeds because it takes the informal, human-centered idea of "effective" – the idea of a human being working with a formal system, or calculating according to a definite rule, in a social context – as a starting point (§4.3, §4.6).[49]

Logic, calculation, and proof depend upon the fact that we *mean*, and colloquial language must be respected because of the fundamental importance of communicability. The TLP admitted this, but suggested that what makes this understanding possible in general must be formally expressible. After he surrendered this "must," Wittgenstein had to face the question of how far we have an informal understanding of colloquial language.

Middle Wittgenstein remarked (PG, p. 124)

> I would like to say: I must *begin* with the distinction between sense and nonsense. Nothing is possible prior to that. I can't give it a foundation.[50]

---

47 Wittgenstein and Waismann, 2003, Preface; Waismann, 1936b, 1936a, 1976/1997, 1982; Shanker, 1987; Makovec and Shapiro, 2019; Shapiro, 2018.
48 On "Church's Thesis" that every computable function is Turing computable see Herken, 1988 and Olszewski, Wolenski, and Janusz, 2006.
49 Sieg, 2009; Floyd, 2017.
50 Wittgenstein, 1974.

This general point about sense and nonsense is crucial for logic, and hence, for mathematics. It is crucial both for formal theories and for informal ones, whether considered separately or in interplay with one another. When we set up a proof, it must be specified, in an intuitive sense, whether or not one *has* a proof. "In an intuitive sense" implies *effectively*, that is, subject to the requirement that one can test whether or not what one has *is* a proof by going over the series of symbolic articulations and following the reasoning step by step, without the admixture of external opinion or controversial principles, in a finite number of steps. This is the computational *aspect* of proof.[51] (It applies also to calculations.)

If such a terminating routine is not available, the proposer of the proof, in any given instance, may reasonably be asked to give a proof that this *is* a proof, in the sense of a forceful ground. There must be what (Turing's dissertation advisor) Church (1932) called a grasp of "intuitive logic": a sense of cogency and reasonableness in appeals to what is obvious and/or relevant. One cannot learn formal logic without "intuitive logic," for one would not be able to understand the instructions for formulating and reasoning *in* formal logic without it.

Ordinarily we grant that a (meaningful) metalanguage may be used to set out a formal criterion of proof. A formal system, insofar as it answers to meaningful aspects of the metalanguage, may be considered a *language*. But hierarchies of formal languages are from Early and Middle Wittgenstein's point of view just more construction by us *in* a unified language. Middle Wittgenstein was influenced by Hilbert's formalism about "foundations," reworking the analogy between formal theories and chess. But he regarded Hilbert's "metamathematics" as "just more mathematics," and therefore not part of a general philosophical answer to fundamental concerns about meaning and justification (WVC, 136; §3.6).[52]

At issue for the Middle Wittgenstein is whether the computational aspect of proof exhausts mathematical meaning (i.e., the rules in an autonomous system). He holds that it does, though the algorithms surrounding mathematical calculations and proofs are localized to independent *Satzsysteme*. He surrendered this idea after 1934, for very good reasons. But it took him a long time – until 1944 – to work out the consequences for philosophy of mathematics. What was in the Middle Period a *requirement* for the meaningfulness of a mathematical sentence (that it be decideable, i.e., proveable) became only one partial *aspect* of mathematical sentences in the Later period.

---

[51] Compare Whitehead and Russell, 1910, Summaries of Part I, *4.
[52] Mühlhölzer, 2012.

Middle Wittgenstein began to respect the complexities of "ordinary language." Simplicity was relativized to *Satzsysteme*. But within each system there still were (realtively absolute) simples. This hybrid account resulted in an unstable mix.

3.2 *Satzsysteme* and Statements of Gradation

Ramsey (1923) forwarded a number of penetrating criticisms of TLP, especially its treatment of arithmetic and number. In RLF (1929) Wittgenstein responded by making concrete proposals, publicly retracting the TLP claim that the GFS alone expresses the formal unity of language and logic.

Ascriptions of graded properties ("It is 80° outside") immediately exclude all other statements of the same degree ("It is 90° outside"). An object cannot simultaneously be both red and green all over. Wittgenstein now states that these are "complete" descriptions of phenomena requiring no further analysis. TLP's structuralizing approach to numbers as operational exponents of purely "formal" possibility forced a distinction between the variables used in different ascriptions of units to situations. "It is 80° outside" should be equivalent to "It is $(70 + 10)°$ outside." But "It is 70° outside" might use "$v_1, v_2, v_3 \ldots v_{70}$" and "It is 10° outside" "$v_{71}, v_{72}, \ldots v_{81}$". To combine the statements would require the final analysis, where all names have distinct references and the ultimate ground of indexing by variables is exposed. Additive possibilities would be explained by "hidden" conjunctions, exclusions by "hidden" disjunctions (possibly infinite ones, given the need for functions of real numbers in physics).[53]

Yet we see right away, on the surface of meaning, so to speak, that its being 80° outside excludes its being 70° outside, that a's being red all over excludes its being green, and so on with all statements of degree and gradation. This shows that the TLP's grammar allows for nonsense, promising to handle it later. Wittgenstein now refuses to put the problem off, taking colloquial language as a touchstone. The shift from hidden (truth-functional) contradictions to nonsense indicates a striking shift. But he still initially proposed the construction of a specialized "phenomenological" language to show *why* the claim that "A is green all over and A is red all over at the same time" is nonsense.

Wittgenstein states that we "meet" in experience "the whole manifold of spatial and temporal objects, as colors, sounds, etc. etc. with their gradations, continuous transitions, and combinations in various proportions." But in statements of degree "we cannot seize …by our ordinary means of expression"

---

[53] Ricketts, 2014 interprets TLP 6.3751.

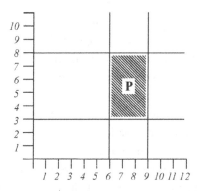

Figure 8 Numbers immediately structure elementary sentences
Source: RLF, 34.

these multiplicities (RLF, 33). Statements involving gradations must therefore involve more than simple names in combination: "numbers (rational and irrational) must enter into the structure of the [elementary sentences] themselves" (RLF, 33). Mathematical form is no longer distinct from structure and internal relations among all sentences. Now the "elementary sentences" are not always formally independent. The GFS is (at least) incomplete.

The outcome of analysis will be *systems* of sentences (*Satzsysteme*). These are held together by internal relations that are not only truth-functional. Quantifiers do not abbreviate infinite conjunctions and disjunctions (MWL 6b:10, 6b:27). Infinity is an infinite *possibility* in the grammar of particular *Satzsysteme*. For example, to say that a patch P is red in the visual field we would write "for a continuous interval", "[6–9, 3–8] R", utilizing the coordinate system as an unanalyzable feature of the mode of expression (RLF, 34) (Figure 8).

Thus Wittgenstein sought to retain what he could of the TLP view. "Grammar is arbitrary" in the sense that there is no universal, formal structure it *must* have (PR §2): there is a variegated archipelago of determinate possible *Satzsysteme*. We must analyze this without *a priori* assumptions or hypotheses. In one system we might take the three primary colors as "simple," and in another, based on the color octahedron, four. In one system of number the rationals are ultimate, in another the reals. The challenge is to find the right multiplicity within this or that region of grammar and state it in rules.

But how was mathematics' universal applicability across *Satzsysteme* to be understood? What about the relations among mathematical systems of number? Once he had given up the GFS, Wittgenstein could no longer attach his account of number to logic directly via a general theory of operations. His "nebulous" introduction of number in the TLP needed rectification (PR §109). He retained a sharp distinction between the purely "formal" activity of ascribing a number to a mathematical structure – he now thought of this as a "purely grammatical" activity – and the ascription of number via material concepts (PR §113): arithmetic is autonomous but universally applicable; like geometry it is developed independently of how the world is.

Wittgenstein first attempted to construct a "phenomenological language" to directly describe both what experience delivers to us and its mathematical multiplicity. But this project soon collapsed.[54]

On the one hand, there were separate spaces (e.g., the "feeling space," the "visual space"). But mathematical multiplicities (e.g., direction) must be absolute *within* each space, or else the spaces could be correlated or merged into one grammatical space. If we are to take ourselves to be able to describe *reality* (i.e., all possible things that may be said, true or false), there must be a way to describe the grammar of each "space." Thus in visual space there is an "absolute" right and left, above and below, magnitude of length, and so on.

For example, we can apparently meaningfully speak in visual space of a part of a red patch, or a yellow patch larger than an orange one. We may say, for example, that two stars alternately appear and disappear at the same *distance* from the border of a visual field. But we cannot say what *really* appears to us. For we have no way of making sense of being wrong here: it is absurd to suppose that we learn sameness of distance in the visual field by correlating our experiences with some external system of measurement (say, a ruler). "The system of coordinates is contained [internally] in the nature [grammar] of the [visual] space" (MS 105, 35).

But now, Wittgenstein soon reasoned, the idea of the "reality" of the visual space has not been given content. It doesn't make sense to say that something can *appear* differently than it *is* in visual space (MS 107, 29). We require *another* space to make sense of this. But this will either be "hypothetical," importing the grammar of physical or other spaces, or it is wholly arbitrarily imposed.

The difficulty lies in Wittgenstein's having imposed the TLP's demand for sharpness of sense on the grammar of visual space. For that grammar, as other possible grammars of experience, is essentially inexact. For me to assign, for example, the coordinate unit to the length of a line in visual space, I must be clear on the difference between a line of length n and $n + 1$, or n and $n + 1.001$. But we aren't able to *see* this (MS 107, 71). Wittgenstein considers Hjelmlev's axiomatization of "rough" visual space and rejects its account's distance from ordinary language (WVC, 35, n. 15). He sees that we inevitably draw on

---

[54] Engelmann, 2013.

concepts from physical language, which have a certain precision, to describe sense data.

The failure of Wittgenstein's phenomenological language drove him back to reconsider the role of ordinary language. In applied geometry, he conceded that we give preference to Euclidean geometry as part of our grammar in relation to what we see (WVC 59). Sentences are embedded in systems of possible experience, he now thinks, as hypotheses that have possible verifications. They are tools for guiding and building expectations and the fulfillment of intentions and actions, which are internally related to the statement of desires and wishes, as well as beliefs and statements.

Differing *Satzsysteme* (e.g., physical and memory time) must not have their grammars mixed, lest we confuse ourselves with nonsense. Internally to a system, questions make sense only as searches whose conditions of fulfillment are clear in advance, and for which there is a definite method of resolution; *how* we search for an answer tells us precisely *which* question is asked. If there is no definite possible resolution of the question *in* the grammatical system, it is nonsense and there can be no search for an answer (PR §§33, 43).

3.3 Aspects: Sheffer's Strokes

The image of wholly independent, equiprimordial *Satzsysteme* raised the issue of what kinds of relations such grammatical systems could have to one another. In Sheffer's stroke – the fundamental inspiration for the GFS – Wittgenstein found the answer. He reinterpreted Sheffer's "discovery" of the stroke as the "discovery" of a new possibility in grammar, a new *Satzsystem*, something of the right multiplicity – which, on Wittgenstein's Middle view, meant the invention of something new (BT 419):

> What was it that we didn't know before [Sheffer's] discovery?
> You can see that very clearly when you think of the objection raised: $p|p$ is not what $\neg p$ says. The answer is, of course, that it is only a matter of the system $p|p$ etc. having the necessary multiplicity. So Sheffer has found a symbolic system that has the necessary multiplicity.

Here we see a signature idea of the Middle and Later Wittgenstein emerging. What Sheffer "discovers" is not merely a trivial abbreviation; abbreviations we can get along without (BT 469). Rather, he introduces a new *aspect* of logic and mathematics: the fact that the logical constants display the "multiplicity" of a Boolean Algebra.[55] Sheffer gets us "acquainted" with a way of regarding things, a new conceptual structuring, an "internal" relation between systems of

---

[55] Sheffer, 1913.

grammar that is not logical in character, but yields mathematical insight (PR §104, BT 478). (Russell took this to be the real value of Sheffer's work, and recommended that he rewrite all of *Principia*.[56])

Wittgenstein's philosophy of mathematics turns on generalizing a sharp distinction between definitions that merely abbreviate (tautology-like substitutive rules for expressions, e.g., the fact that the Sheffer stroke may be used as an abbreviation device in truth-functional logic) and conceptual remodelings that reveal new and fruitful aspects of mathematical structure, re-analyzing, re-dimensionalizing, re-scaffolding, and re-projecting mathematical concepts. He takes these aspectual shifts to be what Kant had aimed at in calling mathematics "synthetic *a priori*" (PR §108).

Wittgenstein includes typical cases (the embedding of the natural numbers in the integers, the rational numbers in the reals), but also the embedding of particular calculations in recursive proofs (BT 469) and even the square of a binomial:

> It's clear that the discovery of Sheffer's system in $\sim p\,\&\,\sim p\ =\sim p$ and $\sim(\sim p\,\&\,\sim p)\,\&\,\sim(\sim q\,\&\,\sim q)\ =\ p\,\&q$ corresponds to the discovery that $x^2 + ax + \frac{a^2}{4}$ is a special case of $a^2 + 2ab + b^2$ (BT 478).

This sense of being a "special case" is not quantificational but aspectual: a deep indication of Wittgenstein's non-extensionalism. He remarks that though we are inclined to regard the ratio $\frac{1}{3}$ as if it *consists* in the extension $0.33\overline{3}$, what we have is more than a mere abbreviation; it is a new a rule of proceeding, exhibiting internal necessities (BT 466). This "face of necessity" is lost if we try to conceive of writing out a real number expansion extensionally; this is like writing down a sentence by copying letters off of coded outcomes of rolling the dice. Sentences have aspects; they are not mere sequences of signs (PR §17, BT 6). The variable does not merely abbreviate, any more than $(\exists x)f(x)$ abbreviates a disjunction (BT 230).

This means that certain examples of "defined" real numbers don't count as aspectually "synthetic." Wittgenstein struggles with this. The number defined from the expansion of $\sqrt{2}$ by substituting "5" everywhere the digit "3" appears is just an "analytic" variant of $\sqrt{2}$ (PR §183). The *general* idea of "number generated from $\sqrt{2}$ by such a transposition," in belonging to no system of measurement or otherwise, is tantamount to $\sqrt{2}$ all over again (PR §183). π', defined as replacing every occurrence of the digit "7" in the expansion of π, is "homeless" until we have a handle on the occurrences of "7" (PR §186). "$\frac{2\to 5}{\sqrt{2}}$" might,

---
[56] Whitehead and Russell, 1910, xv; Floyd, 2021.

for all we know, mean $\sqrt{5}$. The substitution rule $(1 \rightarrow 5)$ "strikes at the heart" of the expansion 0.101001000 …, but how?

Although *Satzsysteme* are internally structured grammars, it is essential that we *can* sometimes make one out *in* another: we can project internal relations across different fields of expressivity. For this, however, they must be distinct conceptual spaces:

> Is it a search, if I don't know the Sheffer's system and say I want to construct a system with only one logical constant. No! The systems are not in one space at all so that I could say: There are systems with 3 and 2 logical constants, and now I am trying to reduce the number of constants in the same way. There is no same way here! (MS 108, 104–107).

Wittgenstein's insistence on the completeness of grammar within systems, coupled with the independence of *Satzsysteme* from one another, inflicts an epistemological constraint. There is no step-by-step procedure or method to reach one *Satzsystem* from another, nor is there a way to "reach" something inside a *Satzsystem* other than calculation (because anything that may be said is already there in the grammar). Therefore, there is no "route," either from one system to another or from one sentence to another within a system. Grammar is modally static and determines a total space. Whatever procedural elements there are are operational, or formal, in the TLP sense. These do not have the power to alter the internal structuring of the concepts.

Wittgenstein has a particular way of getting to a concept of "reality" at this point. There *is* something to see: possibilities are revealed through analysis, discovered. But there is no grammar of "searching" for aspects or proofs. The TLP's sharp distinction between calculation and experiment, formal and material, holds fast. Justification is *constituted* by grammatical rules that are always internal to the *Satzsystem* they form.

It follows that there can be no meaningful mathematical conjectures (PR §148): until proved, a mathematical sentence formulated in ordinary language has no sharp sense. Unproved conjectures are merely linguistic stimuli.

Within this form of verification, mathematical sentences *have* a sense, at least once proved. In the TLP they did not. Wittgenstein has broached a problem he will not resolve until much later: What is mathematical understanding? What is it to grasp the sense of a mathematical proposition or a mathematical concept?

Middle Wittgenstein has begun to put together a form of "realism" with which he can live. In characterizing such-and-such *as* so-and-so we draw out an *aspect* or way in which things *are*. Objects and properties are not collections of aspects; they have internal conceptual (modal) structure. Aspects are there to *be* seen: they are discovered, and so not "merely subjective," purely epistemic, or conventional in character.

> What was it we didn't know [*wussten*] before [Sheffer's] discovery? (It wasn't anything that we didn't know [*wussten*]; it was something with which we weren't acquainted [*kannten*].) (BT, 419)

Wittgenstein contrasts "knowing" that a particular proposition is true with "knowing" in the sense of "being acquainted with": his replacement for Russell's notion of "acquaintance" with forms (§1.2).

That aspects represent the *discovery* aspect of mathematics remains central to Wittgenstein's discussions from this point on. He does not fail to emphasize the "invention" aspect: if he has to choose, he emphasizes this (RFM I §168, I Appendix II §2). In the Middle Period, however, the invention side of things is left hanging, described as a matter of "intuition" or "insight" or "decision" (PR §104,174). Given Wittgenstein's calculus conception of the grammar of *Satzsysteme*, there is no room for conceptual ("grammatical") "discovery" or "invention" within a system. Thus there are no criteria for distinguishing the appearance of "the same" mathematical sentence in two different *Satzsysteme*. Every sentence with sense requires a grammatical place in *some* such system. But how to tell where to place a given calculation?

This broaches the so-called rule-following "problem" (§4.2). It takes several different forms in PR.

First, if we imagine calculating with a decimal expansion of an irrational number there is no periodicity. So even if we have a "rule" according to which we *are* calculating – e.g., digits of π – this may be a "rule" in a different sense from one for a rational number, especially when we consider π in a differ-ent way, say, as a "law" for measuring. The individuality of the expansion's home(s) outstrips what we can say in general about a pattern.

This is a problem in other kinds of cases, such as instantiations of recursive rules with particular natural numbers. Seeing a substitution *as* a correct substi-tution instance is different from the enunciation of a rule *of* substitution such as an equation or logical inference rule (WVC, 154f.). Wittgenstein is tempted at one point to say that there is a new "insight" at each step of a calculation (PR 198), though he draws back from the phrase "direct insight," replacing it with "decision" (PR 129, 171): a collapse into conventionalism. He has seen that there must be a way of grasping a projection that is not an interpretation (from outside the system), but a way of reading (structuring) the signs, and he sees that this cannot be a particular mental state (PG §9).[57] But aspect-shifting

---

57 Fogelin, 2009, 15ff.

sits uncomfortably with the idea of being wholly inside or wholly outside a system.

Second, it is not clear what might allow one to transcend the limits of a particular *Satzsystem*. No room may be made for this, as there was for a series of forms or iterated types in the sense of TLP. Our above diagonal arguments (§2.6) press upon Wittgenstein the problem that "being the same" is neither relative nor absolute across his variety of systems.

3.4 Skolem Arithmetic and the Uniqueness Rule

In TLP the whole edifice of arithmetic depended upon a fixed, absolute multiplicity. What was Middle Wittgenstein to say about his characterization of cardinal number by way of the formal series $[0, \xi, \xi + 1]$? What was to ensure that in taking a step "one more further" in one *Satzsystem* one would still be taking one step further in that system?

Unlike the grammar of visual space, number admits of no vaguenesses. Wittgenstein remarks that a number is a "picture of an extension," by which he means that it is a norm against which an extension is measured within the context of a proposition as a whole. As in the TLP, the *possibility* of (making sense) of a correlation amongst objects and numbers *procedurally* (non-extensionally) is at issue in counting, not any actual (extensional) correlation between objects or signs (PR §100). But if Wittgenstein's modal, non-extensional treatment of sameness of cardinality was to hold up, something had to be universal *in* the idea of counting when it is transposed into the keys of differing *Satzsysteme*.

Wittgenstein read Skolem's (1923) paper presenting a quantifier-free version of arithmetic. This stimulated him to reformulate his reliance on what we earlier called (§2.5) the Uniformity Principle: Poincaré's idea that a "uniform" operation applied to "the same" inputs will yield the same result. Associativity of arithmetical laws was grounded in the GFO in TLP. Now Wittgenstein presents the generality of associativity as a "basic rule of the system" of arithmetic, namely a grammatical, constitutive rule *about* which nothing can be proved, because it enunciates an infinite possibility of application to particular numbers (PR §163).

The same status seems to be attributed to the equality of a number with itself. Wittgenstein asks why one cannot deny a basic rule of the system such "2=2", but can deny an equation such as $2 \times 35 = 70$ (PR §163). There are aspects of the grammar of an equation that one can miss, so to speak. Yet miscalculations are capable of being shown not to "fit" the grammar of arithmetic.

In the TLP the natural numbers were "all" built in to the general form of form series. Now Wittgenstein criticized Skolem's failing to make clear that "all" must already reside in the very ideas of an operation, a form series and a formal substitution: "all" numbers must be always already "seen *in*" the "calculus" of (quantifier-free) arithmetic. Skolem's recursive proofs yield, not propositions with sense, but "signposts," guides pointing us toward specific steps we *can* take with it (PR §163).

Skolem's treatment of the associativity of addition, key to his step-by-step substitution of terms in the quantifier-free setting, failed to draw out this modal character. For Wittgenstein, Skolem's failure to acknowledge his reliance on the Poincaréan, modalized Uniformity Principle, a fundamentally non-extensional, *procedural* sense of rule, was lacking.[58]

"Every symbol is what it is and not another symbol," remarks Wittgenstein, alluding to Bishop Butler (PR §163). When one chases down the implications of an inequation, one reaches nonsense. This suggests a connection between a procedurally conceptualized version of the Pigeonhole Principle (if one has n items and m boxes, $n > m$, one cannot put items into the boxes without putting more than one into at least one box) and the Principle of Mathematical Induction. Such a connection there is: mathematical induction proves the Pigeonhole principle, which therefore appears as conceptual, something about procedures. (This does not prevent one from regarding it under the aspect of an empirical fact.)

Wittgenstein works with the idea that the natural number instances of the laws Skolem investigates serve as *parameters*. The question emerging is, With what right can we operate with parameters at all in mathematics? How can a procedure of shuffling parameters *show* the generality of a theorem? Once again: the Uniformity Principle is logically fundamental.

Ordinarily one regards elementary, quantifier-free arithmetic as a schematic template for reasoning hypothetically about characterizations of the numbers. Or one quantifies second order over properties. As we have seen, neither route is acceptable to Wittgenstein for constructing a primordial "foundation" of arithmetic.

R. L. Goodstein, who later made a number of contributions to logic and recursive function theory, attended Wittgenstein's Cambridge lectures and his dictations (BBB). On May 20, 1932 – as Marion and Okada (2018) have shown – Wittgenstein formulated a rule asserting the uniqueness of a function defined by primitive recursion. For arbitrary terms $u(x), v(x), w(x,y)$, and $S(x)$ the successor function, the rule may be stated, without parameters, thus:

$$\frac{u(0) = v(0) \quad u(S(x)) = w(x, u(x)) \quad v(S(x)) = w(x, v(x))}{u(x) = v(x)}.$$

---

[58] PR XIV, PG , MWL, PG 397ff., BT 445ff. – Compare, however, Goldfarb, 2018a.

The rule matters to Wittgenstein's Middle philosophy because it reinforces the idea of a widely transportable, minimal formulation of arithmetic characterization in terms of quantifier-free, step-by-step operations expressing the Uniformity Principle. Goodstein appears to have learned of this rule from Wittgenstein, later proving that it implies mathematical induction for primitive recursive arithmetic.[59]

Goodstein also developed a "bump" function leading to the characterization of a function whose arithmetical complexity outstrips what quantified arithmetic can decide. This was the first "natural" example of an undecidable sentence after Gödel's (1931).[60]

What Skolem's recursive proof indicates is the possibility of plugging in any instance of a number one likes. One "sees" that "the proposition is a member of the series of propositions that the final proposition of Skolem's chain presents." Wittgenstein holds that this form of recognition is "not provable, but intuitive" (PR §163). But then there appears to be no ground for the idea that the Uniformity Principle is a *general* "aspect" of all such intuition: what work the quantifiers do, as we would say today, is intrinsically more complicated.

Two nascent ideas begin to emerge.

First, the underlying question for the Uniformity Principle is "What is it to apply the same rule *to the same thing?*" Wittgenstein has already admitted that there is a way of applying a rule to an instance that is not an interpretation, but is simply exhibited, projected, seen. If there were constantly clarifying interpretations all the way down, there would be no end. The Uniformity Principle depends upon that admission.

Second, Wittgenstein's modal rendition of the Uniformity Principle does not suffice to spell out what a number, in general, *is*. If the principle of mathematical induction, regarded quantificationally, is not built *in* to our concept of number, then there is a seeming gap about which numbers there *are*, and what holds their aspects together. The "internal relation" among proof procedures and systems that we should be able to see, in a way analogous to what we saw in the case of Sheffer's projection of his stroke from the concept of a Boolean Algebra, is lacking.

According to Goodstein (1957a),

> Wittgenstein with remarkable insight said in the early thirties that Gödel's results showed that the notion of finite cardinal could not be expressed in an axiomatic system and that formal number variables must necessarily take

[59] Goodstein, 1957b; Marion and Okada, 2018 gives details.
[60] Paris and Harrington, 1977, https://mathworld.wolfram.com/GoodsteinSequence.html.

values other than natural numbers; a view which, following Skolem's 1934
publication, of which Wittgenstein was unaware, is now generally accepted.

3.5 Real Numbers: the Calculus Conception

At BT 738 we read:

"The rational points lie close together on the number line": a misleading
picture. [...]
    Is a space conceivable that contains only all rational points but not the
irrational points? And that only means: Aren't the irrational numbers prejudged in the rational ones? No more than chess is prejudged in draughts.
The irrational numbers do not fill a gap that the rational ones leave open.

This attack on "gap" imagery not only advocates the idea of separate *Satzsysteme* of numbers. It also rejects the extensionalist perspective that permeates
modern presentations of the real numbers. Wittgenstein's game-comparison,
ridiculous on its face, is a therapeutic manoeuvre. He is inviting the reader to
see the relationship between rational and irrational numbers in terms of aspects:
not as one of *extending* a domain, or providing an adequate extension for the
concept of the number line or space, but rather as one of conceptual *embedding*,
the structuring of one *kind* of number in terms of another.

    Wittgenstein encountered the language of "gaps" in Russell (1903, §§272,
343) where the theory of limits and continuity are treated (compare Russell
[1920, Chapter X]). The "gap" rhetoric is common in many other texts, starting
with Dedekind (1872, 771f.) and continuing with Waismann (1936a, Chapter
8); it pervades texts down to today, even textbooks of pure set theory (Hrbacek
& Jech, 1999, 86f.). Dedekind's analysis of real numbers as "cuts," or sets of
rationals, offers a satisfying way of closing off the real numbers.

    But the point of these texts is to leave *behind* the geometrical illustration,
replacing it with a purely arithmetical conception of real numbers in which the
reals are wholly divorced from magnitudes. Like Frege, Wittgenstein resisted
this severing of an account of the real numbers from ratios (§2.4). Wittgenstein has no problem with *individual* irrational numbers like $\sqrt{2}$ and π when
these are determined by definite techniques of producing their decimal expansion or similar means of identification. He balks, however, at the extensionalist
transition from the rational to "all" real numbers at one stroke.

    Weyl criticized the impredicativity of the Dedekindian conception (1918),
and Brouwer forwarded intuitionism. Wittgenstein knew enough of Weyl's
and Brouwer's works to formulate responses.[61] There is a whole program in

---

[61] Marion, 1998, 2003.

predicative analysis.[62] And even mathematicians who do not advocate predicativity or constructivism sometimes raise the question concerning the real justification of the transition from rational to real numbers.[63]

Wittgenstein distinguishes sequences of numbers that the extensionalist considers to be, in Cantor's sense, "finished" (*fertig*) – these are the "extensions" – from the techniques or rules of development by means of which such entities may be produced, assessed, or individuated: let us call these "expansions." According to his non-extensional way of looking at things, the decimal expansion for $\frac{1}{3}$, $0.\overline{333}\ldots$, is regarded as a technique for developing digits. If we are interested in *particular* real numbers (like π), we focus on such techniques. Some may be purely mathematical. Others may be shaped by geometrical considerations or applications in physics.

By contrast, the extensionalist's interest is only in the results, the produced sequences (let us call them, speaking extensionally), and not the possible processes or conceptual motifs or techniques leading to them. From the extensional point of view, expansions are mere "illustrations": perhaps useful to know but inessential (WH §5.5).

We may take what we regard as a purely arithmetical (i.e., nongeometric) result – e.g., that $\sqrt{2}$ is irrational – to "apply" to geometry. Transposing onto the straight line, we accept, for example, that the diagonal of a unit square does not end in a rational point. We have a procedure "determining" a particular point in a few straightforward geometrical steps. This has an altogether different aspect from the "procedure" of Dedekind, which "determines" a point by approaching it from below and from above via a "procedure" consisting of infinitely many steps. Considered as an "application" of analysis, this obviously does not come to an end.

There is a characteristic tension in our concepts, as Bernays (1957, 4) remarks:

> The conflicting aspects of the concepts to be determined [for analysis] are, on the one hand, the intended homogeneity of the idea of the continuum and, on the other hand, the requirement of conceptual distinctness of the measures of magnitudes. From an arithmetical point of view, every element of the number sequence is an individual with its very specific properties; from a geometric point of view we have here only the succession of repeating similar things. The task of formulating a theory of the continuum is not simply descriptive, but a reconciliation of two diverging tendencies.

---

[62] Feferman, 2005.
[63] Gowers, 2017a.

This sort of tension is insisted upon in Wittgenstein's remarks.[64] Even Hardy (1941, §16), critically annotated by Wittgenstein (WH, §4.5), is sensitive to the aspectual "conflict": "We may think of [the real numbers] as an *aggregate* [...] or *individually*."

But the Middle Wittgenstein cannot live with tensions (BT 763ff):

> "The position of all primes must somehow be predetermined. We just figure them out successively, but they are all already determined. God, as it were, knows them all. And yet for all that it seems possible that they have not been determined by a law." – Time and again there's this picture of the meaning of a word as a full box, whose contents are brought to us along with the box and packed up in it, and now all we have to do is examine them. ...We use the expression "prime numbers", and it sounds similar to "cardinal numbers", "square numbers", "even numbers", etc. So we think it can be used in a similar way, but forget that for the expression "prime numbers" we have given quite different rules – rules *different in kind* – and now we get into a strange conflict with ourselves.

The "strange" conflict is that we adopt the extensional viewpoint and yet simultaneously lean on the variety of procedures that give the real numbers their interest for us. For Wittgenstein, there is no fudging the perspectives. And this is an important logical point.

Considered in Wittgenstein's way, the differing domains of kinds of number are "the logically most different structures, without any clear limits" (BT, 751). Even the irrational numbers are, among themselves, "different kinds of numbers – as different from each other as the rational numbers are different from each of these kinds" (BT, 761f.). This is a consequence of the *Satzsystem* conception, which fixes grammar of number words completely *inside* a system.

The irrational numbers may, however, legitimately be regarded as being of different "kinds." Mathematicians distinguish "measures of irrationality" and "hierarchies of complexity" for real numbers.[65] Moreover, we may distinguish kinds of geometrical concepts. Wittgenstein's favorite example is the proof that it is impossible to trisect an angle with ruler and compass alone: the proof projects the concept of what it *is* to so construct a figure *in* to higher algebra, shifting its aspect.[66]

But Middle Wittgenstein's way of making these points costs him. He does not want to "wipe out the existence of mathematical problems" in general, or

---

[64] Bernays, 1959: Wittgenstein's "main argument" in RFM offers an applicable criticism in cases where the intensional and extensional approaches are mixed, creating "the impression of a stronger character of the procedure than is actually achieved."

[65] http://mathoverflow.net/questions/53724/are-some-numbers-more-irrational-than-others.

[66] Floyd, 1995.

hold with Weyl that "the opposite" of a proved sentence "never had sense" (PR §148). But his *Satzsystem* conception imbibes from the TLP the idea of sharpness of sense, putting it together with an idea of mathematical sentences as embedded in calculi that determine their senses. The result *does* wipe out the meaningfulness of unproved conjectures. Wittgenstein also thinks that the idea of the undecidability of a sentence in a system *requires* the extensional point of view (PR §174), something he later doubts (§4.6).

The absence of anything to be said about the transition from not understanding a sense to understanding it raises the spectre of Wittgenstein's Later concern with rule-following (§4.2). Aspects and insights are either rigidly fixed inside a calculus, or else they run wild (PR §148):

> Is it like this: I need a new insight at each step in the proof? This is connected with the question of the individuality of each number. Supposing there is to be a certain general rule (therefore one containing a variable), I must recognize each time afresh that this rule may be applied *here*. No act of foresight can absolve me from this act of *insight*.

This makes it sound as if mathematics is everywhere permeated by intuitions. The idea of a "calculation" exhibiting any necessity, any sense of a "law," then disappears. But this compromises Wittgenstein's ability to attend to differences in perspectives, because this is no better than the extensionalism he resists. He tries reformulating, saying that "a mathematical proposition is a pointer to an insight" (PR §174). But the pointing doesn't point anywhere in particular. So it cannot be pointing.

Mathematics as an activity of "calculation": everything *about* calculi is mere "prose," which has nothing to do with the workings of proof. But Wittgenstein's calculus-relativism entangles him in a myth of symbolism. "Set theory builds on nonsense" and is a "fictitious symbolism" that is "wrong" "because it apparently presupposes a symbolism which doesn't exist instead of one that does exist (is alone possible)" (PR, §173f.) What is "alone possible" is this (BT 468):

> In mathematics *everything* is algorithm, and *nothing* meaning [*Bedeutung*], even when it doesn't look like that because we seem to be using *words* to talk *about* mathematical things. Even these words are used to construct an algorithm.

Yet the limitative results of Gödel, Turing, and others would soon show that this claim is not comprehensive, since it applies neither to the language of arithmetic nor to first-order validity. There is also the nascent rule-following difficulty: *How* do words get "used to construct an algorithm"?

For Wittgenstein, the "fairy tale" aspect of set theory resides in its having already left behind the rules and procedures we use to discuss and apply real numbers as magnitudes, our "calculus" (BT 468):

> What a geometrical proposition means, what kind of generality it has, is something that must show itself when we see how it is applied. For even if someone succeeded in meaning something intangible by it [e.g., Dedekind with his cuts] it wouldn't help him, because he can only apply it in a way which is quite open and intelligible to everyone.
> Similarly, if someone imagined the chess king as something mystical it wouldn't worry us since he can only move him on the 8 x 8 squares of the chess board.

But this dogmatic idea of an application being "quite open and intelligible to everyone" in contrast to its "mystical" interpretation is marred by Middle Wittgenstein's calculus conception. Once he surrenders it, the line between calculation and experiment becomes occasion-sensitive, and he must adopt a more critical anthropological standpoint on what it *means* to be "open and intelligible to everyone" (§§4.1, 4.2).

3.6 Consistency

Middle Wittgenstein's non-extensionalism implies that there can be no central epistemological problem about consistency (WVC 122, 196). A *Satzsystem* is not an extensionally conceived set of statements, but a collection of procedures held together by a set of grammatical rules that can (always) be unearthed and altered. There has been much scandalized response to Wittgenstein's remarks on contradiction, which are revisited in his Later Philosophy. But his remarks, even in the Middle Period, are not incomprehensible or uninteresting.

Though Marion and Okada (2018) do not differentiate the Later from the Middle Wittgenstein, as we do, they offer a sound review of the literature and useful responses to Wittgenstein's critics. They rightly emphasize two main points.

First, if one runs into a paradox in the "prose," or logical forms used to articulate a piece of mathematics, then one can blame the prose, as happened with the paradox in Frege's system (WVC, 120).

Second, the most one can obtain in a relative consistency proof is that a similarity of aspects, or internal relations, may be revealed between two calculi. That is not to make one or the other of the systems a primary "foundation." As to "absolute" consistency proofs, it seems Wittgenstein never worried too much about this. If obtained, they would not speak to Hilbert's occasional construals of possible inconsistencies in systems as cancerous growths that may explode, attacking the body of mathematics from within. This is too "indefinite" a fear (RFM III §86): models are not given by *one* method (WVC, 148f.). Finally, it is not clear the discovery of a paradox need not vitiate all that came before. An inconsistent system may allow us to pursue alternative paths forward, raising "a very exciting question" (WVC, 197), namely, "What shall we do?"

Here I draw a sharper contrast between Middle Wittgenstein and Later Wittgenstein than most readers do – including Marion and Okada. It is difficult to see how the Middle Wittgenstein would deal with the "excitement" of a paradox. For he cannot make sense of the idea of "altering" the rules of a game: the relevant rules are constitutive *of* the game, not merely strategic, and to alter them is to begin anew. (Chess played without two bishops is not chess.) But then Wittgenstein's impulse to downplay the epistemological weight of consistency proofs cannot be articulated. Moreover, the *interest* of what we *might* do in response is erased from view, since this would be a practical question (hence not something mathematical).

We see some clues as to where Wittgenstein needs to move in his Later Philosophy when we look at what he said, by contrast, in his 1939 response to Turing's objection that the use of an inconsistent system could land us in trouble – say, the collapsing of a bridge (or, nowadays, the collapse of an operating system). Wittgenstein replied (LFM, 209f.):

> Now one can imagine an enormous number of rules or axioms written on an enormous blackboard. Somewhere I have said *p*, and somewhere else I said ~ *p*, and there were so many axioms I didn't notice there was a contradiction.
> Or suppose that there is a contradiction in the statutes of a particular country. There might be a statute that on feast days the vice-president had to sit next to the president, and another statute that he had to sit between two ladies. This contradiction may remain unnoticed for some time, if he is constantly ill on feast-days. But one day a feast comes and he is not ill. Then what do we do? I may say, "We must get rid of this contradiction." All right, but does that vitiate what we did before? Not at all.
> Or suppose that we always acted according to the first rule: he is always put next to the president, and we never notice the other rule. That is all right; the contradiction does not do any harm.

The first thing Wittgenstein imagines is that, through the number of rules, axioms, or inferences he has written down, he fails to properly *survey* the situation: there is no "perspicuous representation" of what is going on. A major theme in his Later remarks on mathematics will be the norm of surveyability, which he connects with the concept of *proof* (§4.4). The second is an anthropologically oriented discussion of how people might actually arrange their lives. His Later point is that words (and mathematical systems) must be embedded in situations, they do not carry themselves forward without our ushering them, and we do this with particular purposes in mind, and for a variety of reasons.

The two points are connected. Later Wittgenstein differs from Middle Wittgenstein insofar as he attends, not to discrete *Satzsysteme* conceived of as internally complete calculi, but to the embedding of these calculi in what he calls, only after 1937, "forms of life." That makes all the difference to his Later remarks on mathematics.

4 Later Philosophy (1937–1951): Fluid Simplicity

Later Wittgenstein exhibits mathematical thinking against the backdrop of an evolving sea of human activity. Invention of mathematical techniques and ongoing informal discussion shape it everywhere. What is evident, obvious, "unthinkable," clear, "simple" still plays a role, but it may be contested and shift. Simplicity is fluid.

Aspect-talk in Wittgenstein's Later Philosophy continues to voice dimensions of possibilities and necessities of articulation, the "sides" or "faces" of numbers, thoughts, situations, concepts, proofs, procedures, and so on. But now aspects appear as part of the unfolding of what Wittgenstein calls *forms of life*: the continual embedding and interweaving of mathematical concepts and signs *in* life. Form is still "the possibility of structure" (TLP 2.033), but now the modalities involve our weaving the carpet, or ribbon, of life (PPF §§2, 362). We can swim with our concepts in an evolving, fluid setting, pursuing syntheses and analyses – if we know how to move our limbs. What we require are mathematical *techniques*.

The terms "technique" and "forms of life" enter Wittgenstein's manuscripts in 1937; he drops the notion of "culture" and introduces his Later interlocutory style.[67] Voices frame arguments and objections, revisiting and varying their articulations over and over again. This marks a sea-shift.

In BBB, to imagine a language is to imagine a "culture," but now it is to imagine a "form of life" (PI §19): a specific modification of the human animal using language to articulate. Forms of life have both a biological and an ethological dimension: they are generic in human forms of life (that we chat, that we calculate) and specific to particular ways of embedding these in particular lives.[68]

Wittgenstein gets significantly beyond his Middle period just here, assembling a distinctive philosophy of mathematics.

---

[67] Floyd, 2018b.
[68] Cavell, 1988; Moyal-Sharrock, 2015.

4.1 Language-Games to Forms of Life

In BBB the notions of *language-game* and *family resemblance* become salient. Criteria of identity fluctuate as the borders and seams between systems are given an anthropological cast. The question that arose in Wittgenstein's Middle Period about the Uniformity Principle, videlicet, "What counts as 'the same'?," is faced head-on. Uses of criteria for concepts reflect mastery of language: not something everywhere bounded by rules. For Later Wittgenstein, this allows for – in fact demands – the continuing invention of new techniques.

"Language-games" are simplified snapshots of portions of possible human linguistic behavior, "objects of comparison" offering localized "logic": aspects of concepts and procedures. Human beings operate with signs; mathematics consists of "what can be written down" (BBB 4,6). But the idea of "direct symbolizing" in the sense of algorithms in *Satzsysteme* disappears.

There is richness and complexity in even "simple" scenarios of articulation: our characterizations evolve as we re-characterize their significance in light of our characterizations. Aspects of language-games mutually interpenetrate in different ways, many of them unforeseeable until subject to further characterizations. Wittgenstein now questions whether a "complete" specification of grammatical rules is useful or even attainable for many concepts.

"Family resemblance" characterizes the generality of certain concepts. A single property, a fixed-for-all-cases criterion, an explicit set of grammatical rules – these are not required. A concept may hold together – like a family – with a variegated, evolving series of properties (PI §§6, 236 suggests the concepts of *number* and *calculator* as examples).

Meaning, proof, and analysis are now "occasion sensitive" phenomena.[69] Declarative statements may be said to express truth-conditions sharply; one may subscribe to bivalence and the law of excluded middle, or one may not. But individuating *which* among the several possible truth-conditions expressible with a sentence-form *is* expressed on a particular occasion is a matter of being able to embed words in specific situations meaningfully.

Wittgenstein does not reject, but turns on its head his earlier idea of a truth-condition, thereby securing and deepening it. He does not shift from taking truth-conditions as primary (TLP) to taking assertion conditions as primary (PI).[70] Rather, in the very name of our concepts of *truth* and *reality*, forms of life and techniques are needed to *fit* our concepts within life, to fill them out

---

[69] Travis, 2006.
[70] As do Dummett, 1959; Wright, 1980; Kripke, 1982; Maddy, 2014.

and allow them their places. This is Wittgenstein's "realistic realism." Plasticity and the "friction" of life are required for truth.

Logic works not by unearthing grammatical rules of complete *Satzsysteme* but by characterizing possible ways in which activities of articulation might proceed against an evolving, dynamic backdrop of discussion, desire, action, and need. There is a thoroughgoing reliance on taken-for-granted, informal parameters in philosophical argument. What is simple is now that which is taken for granted and uncontested, obvious, "the given" (PPF xi §345). This is a norm of elucidation rather than an actuality.

The Later Wittgenstein's idea that a variety of techniques and aspects is crucial to the objectivity of logic and mathematics is not an easy one, and it did not come easily to him. MS 160, 18r-27r (1938) contains a train of thought where he shifts his view. First he tries to associate just *one* technique of applying a sentence to each sentence (or just one application of a description). Eventually he tells himself to "get used to seeing a variety of techniques of sign usage (i.e., of thinking)" *in* a single sentence: he must overcome a "prejudice" he used to hold *about* the notion of a technique (MS 160, 27r): that it was a purely formal or grammatical phenomenon.

Sentences have multiple faces: aspects revealed by differing techniques of their application. Choices confront us in gauging the potentialities for how any given sentence may be employed. This means that Wittgenstein's older concepts of sense, meaning, and truth are "useless" for him (MS 160, 27r).

4.2 Rule-Following: Paradigms and Proofs (RFM I)

Unlike Feyerabend's and Kuhn's Wittgenstein-inspired notions of a "paradigm," the Later Wittgenstein's uses of this notion – ubiquitous in RFM I – are miniaturized, signifiying partial, shiftable viewpoints, not total worldviews. Both Kuhn and Feyerabend regarded the meanings of theoretical terms (and concepts) to be *constituted* by the role played within a whole theory or at least a whole outlook on the world (much like the Middle Wittgenstein regarded "grammatical rules"). This form of theory-ladenness of meaning is rejected by the Later Wittgenstein.

His grounds for this are similar to those offered by Quine and Putnam. The two ideas that (1) a sentence's meaning circumscribes its range of verification sharply (whether it is an empirical sentence or a mathematical one) and (2) that there are purely "analytic" truths, true by convention or in virtue of meaning or stipulation alone, without entanglement in any range of experience, are dogmas.

Putnam (1962) undercut the Feyerabend–Kuhn view by emphasizing that the range of "unthinkables" outside the reach of a conceptual scheme is porous and evolving, liable to recast itself in the light of experience. First, physics may establish that what was once unthinkable, a priori false – e.g., "One can reach the place from which one came by traveling away from it in a straight line and continuing to move in a constant sense" – is in fact *true*. Second, our descriptive judgments are in part normative, relying on deference to experts and community-wide beliefs and practices. Third, concepts are not like functions that determine, for any given object in the universe, whether or not that object falls under the concept: they are not sequestered in thought-systems whose logic is fixed. (Putnam eventually drew from his arguments a rejection of functionalism, a doctrine in philosophy of mind that he himself invented.[71])

Recall that in the Middle Wittgenstein a difficulty arose with the individuality of particular numbers in relation to a schema or rule. What *is* the universal instantiation step from

$$(F(0)\ \&\ \forall x(Fx) \rightarrow F(Sx)) \rightarrow \forall x(Fx), \text{ and } F(0)$$

to
$$F(10)?$$

Wittgenstein always rejected the idea that a quantificational account of this necessity in terms of the "law" of universal instantiation is satisfactory: he casts it instead as a *projection* of the instance *as* an instance of the rule. But now his earlier ideas of "showing" and "substituting in a calculus" are to be unpacked in light of his Later conception of occasion sensitivity. He retains the idea of projecting a concept or number by drawing out a new aspect but widens aspect perception beyond the incorporation or projection of one rule-governed system into another.

Now he focuses on the fact that although the whole possible use of a concept is generally never before us, we still are able to recognize, in our forms of life, what is to count as "the same" and what not: we develop techniques, which have a practical generality and repeatability. Techniques are required to project aspects, and they require investigation (Wittgenstein explores at length "the grammar of 'technique'" (WH Chapter 8); in LFM "technique" occurs 117 times in discussion with Turing). This allowance for plasticity in projecting concepts shows, not that our procedures are *not* rule-governed, but rather that that notion itself requires parochial elements.

Earlier Wittgenstein's "mistake" (PI §188) was to think that unless the grasp of a formal rule pinned down its applications sharply and totally, the necessary generality of logic and mathematics would be impossible. Later Wittgenstein

---

[71] Floyd, 2020.

is neither denying nor radically conventionalizing phenomena of necessity and determination – he sharply distinguishes "forms of life" from a consensus of opinions, and techniques are neither brute stipulations nor social hardenings. Instead he retains his non-extensionalism while rejecting his Earlier idea that "formality" is self-standing.

Invoking Turing's language-game-like construal of a human calculator as a "machine symbolizing its own actions," he criticizes the idea of a logical "machine" whose possible movements are given like an extension (RFM I §123ff., PI §§189ff.). Turing showed that there can be no machine that is able to determine whether or not one sentence follows from another one formally: this resolution of the *Enscheidungsproblem* is a way of mathematically grounding Wittgenstein's point. The "mistake" is to insist that either a formula determines all the possible steps we may take with it (as a fixed criterion) or else it does not determine the steps. Progress is made only when one understands that the notion of a partial function (one not everywhere defined) is the basic, most general notion.

Wittgenstein investigates how we ordinarily actually *use* the idea of the "steps being determined by the formula" in everyday life: how the concept is embedded. This fine-grained exploration of details may seem irrelevant to the metaphysical question, but Later Wittgenstein has deepened his understanding of the Uniformity Principle, taking it now to reside in our ongoing repair and fashioning of forms of life. "Consequent" and "inconsequent" in the sense of "having import for" are not mathematizable notions (MS 162b, 68r).

This works through to the end the question of the status of the Uniformity Principle, the very basis for the generality of conclusions in mathematics and logic. Wittgenstein wants to know wherein the very idea of "doing the same operation over and over again" resides. Wherein lies the necessity "in" the series "+0", which "yields"

$$2, 2, 2, \ldots?$$

Wittgenstein comments that if intuition is needed for any formal series, it is needed also here, in each and every step, just to hold even and do "the same" (RFM I §3). His point is not that intuition is needed, or that language alone covers its needed role, but that attempting to lodge necessity in one form is a mistake: aspects emerge and concepts are projected through a variety of possible techniques.

There is still the distinction between what Wittgenstein earlier thought of as "synthetic" projections, which reassemble concepts, and "tautological" abbreviations that do not. But the distinction is dynamic relative to our training, mathematical knowledge, and everyday regularities in our empirical world.

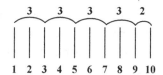

Figure 9 "Miscounting" strokes: $4 \times 3 + 2 = 10$
Source: RFM I §137

If lumps of cheese randomly doubled in size (PI §142) or we could suddenly take in one million parentheses at a glance unaided by a machine, our forms of life – even mathematics itself – would be different.

The fact that we are *not* struck anew by each move, that we have the capacity to *unquestioningly* treat the recurrence of a sign as "the same sign," is as important as our capacity to grasp the shiftings of aspects and projections of our concepts. Wittgenstein has come to see that even in counting the recurrence of a sign as "the same sign" we are exercising our capacity to see one thing in terms of another. But in any such projection, something is unquestioned, passive, taken in. Seeing the duck in the duck-rabbit projects while it reveals, setting up a whole field of necessary relations (if *this* is a beak then *this* is the back of the head, etc. etc.). There is something to *be* seen, in the sense that there is something that may be missed (like missing a solution to a puzzle-piece arrangement). But it is both experience (knowledge of ducks and Gestalt puzzles) and techniques ("here, trace this line as the back of the head, see?") that allow us to see the duck and the rabbit in Jastrow's figure.

What is seen and what is missed may depend upon favors of fate in the empirical world, on previous mathematics, and on our purposes, inventiveness, and interests. A ruler that shrinks to half its length when taken from one room to another might be perfect for a world that behaved oddly in the second room; a "miscount" based on stroke notation yielding $4 \times 3 + 2 = 10$ (Figure 9) might be used to successfully share nuts among 10 people if the distributor gave and took away nuts in accordance with the loops (the analysis) used in the miscount (RFM I §137). What is "unthinkable" in one scenario becomes thinkable, mandatory, in another.

According to Kripke (1982), Wittgenstein was a sceptic about the very idea of *following* (as opposed to making up) a rule, because he establishes that there is "no fact of the matter" as to which projection is correct on a given occasion of use. If we ask a pupil (PI §185) to carry out the calculation of the rule "+2", and at a certain point she "deviates" from what we regard as correct, writing

$$2, 4, 6, 8 \ldots 1004, 1008, 1012 \ldots$$

we have no justification for calling her rule-following "incorrect." For she may answer us by saying that this is how she understood the command all along. In "reality," she meant by "+2" "quus", rather than addition, where the quus rule deviates from addition at a certain point.

On this view we are powerless to convey the necessity of addition through our terminology and/or the exemplification of computations in everyday life, not only because specifying the natural numbers and functions upon them in everyday language requires an ellipsis,

$$0, 1, 2, 3 \ldots$$

but also because, more deeply, the conveying of meaning cannot reach beyond itself into actually uncomputed territory.

Kripke successfully rejects a number of frequently discussed responses, including the idea that somehow speakers are wired up with material dispositions to react in the ways they do, "programmed" by training and/or innateness. He denies that there could be any metaphysical fact constituting meaning one rule rather than another on a given occasion of use. From here Kripke extrapolates to general skepticism about any concept. This is known as the *"rule-following problem,"* a seeming paradox as it appears, quite to the contrary, that in everyday life we manage to convey necessary steps quite well.

Wittgenstein's answer is that there must be a way of following a rule that is not an interpretation or clarification or further analysis, but is exhibited in life (PI §201). Our lives themselves show meaning: something not Humeanly "brute" because it is plastic, normative, and evolving.[72]

Kripke attributes to Wittgenstein an *Ersatz* for logical and mathematical necessity, as well as meaning generally: he takes Wittgenstein to hold that community consensus determines what follows from what, which rule was meant by a given command (e.g., "add 2") – and this is all that can be said: we exclude the wayward rule-follower from the community as a "lunatic" (BrB §30). Smoothness in communication and reasoning is a function of evolving consensus and convention alone: in "fact" it has no *reality*.

But Wittgenstein's point in RFM I and PI is not that consensus is inevitable if we are to have meaning; he is not working with an idea of consensus of opinion at all (LFM 183f.). Rather, and deeper, forms of life weave words and lives together. The idea is that our shared natural reactions, such as they are, the embedding of words and signs as we do it in ordinary forms of life is the "friction" that the very concepts of agreement and disagreement – i.e., truth and falsity – require (PI §§240ff.).

---

[72] Fogelin, 2009, 99 regards the requirement for exhibition in life as Humean and "brute."

The point about "wayward" rule-following is not the difficulty of coming to know "their" mind or ours, but in coming to know our own in relation to theirs: the very idea of a public needs construction before agreement and disagreement, true and false, make sense. I no more possess pearls in a box of meaning through introspection than they do. If each of us adopts the extensional point of view, there is an impasse. Instead, we fashion our ways of meaning together, coming to appreciate different ways of seeing and projecting "the same thing."

If a tribe frequently "came to blows" over whether the letter σ consists of a single etch on the blackboard or repeated occurrences of uses of the sign, they wouldn't count as practicing "mathematics" (PI §240; LFM I). It is not that such coming to blows does not happen. Setting up a code-language always involves struggles. Brouwer and Hilbert came to professional blows over the law of excluded middle in infinite contexts. Nevertheless, the routine communicability of procedures that can be humanly followed is a mark of the logical and the mathematical. At some point the formation of what is "obvious" or "a normal response" is part and parcel of what thinking in logic and mathematics *are*.

In Wittgenstein's later philosophy, this points toward a naturalistic, evolving web of contingencies on which ride the necessities of logic and mathematics. However, it is important that pointing to any particular natural fact will not explain or justify the development of a mathematical technique *as* the technique that it is. For that, we must understand its place in mathematics and hence, in our lives. It is notable that Turing's analysis of taking a "step" in a formal system *assumes without fuss* the living, human context of reckoning according to a rule or technique of calculation.

To count as a technique, there must be something transportable and communicable, something general to be taken and applied in different cases and contexts, adaptable to new contexts, available for tailoring. Mathematicians sometimes speak of a "trick" in a proof: a technique is something else. The generality of techniques is not quantificational, but purposive: techniques are not accidentally or arbitrarily concocted, but are ways of proceeding and projecting with concepts.

RFM I refashions Wittgenstein's calculation versus experiment distinction (§2.5). Now it is viewed as dynamic, embedded by way of techniques in forms of life.

If we take a pile of 100 marbles and separate them into 10 groups of 10, that is an "experiment" in the sense of counting how many marbles are there (RFM I §36ff.). When we begin, we do not know the answer. But if we now film the procedure, regarding it as a paradigm of what *can* always be done, as a *necessary* feature of there being 100 marbles, then we regard the procedure as

a paradigmatic "calculation," or proof: we have "unfolded" the implicit properties in the collection. We "see the necessity" in the procedure: the way of
regarding typical of proof. This is a shift from the aspect of experiment to that
of calculation: from a procedure of discovering in empirical fact about what
may be done with marbles and objects sufficiently like them, to the revealing
of an aspect: a configuration of concepts.

The generality in the second case is not quantificational: we do not infer
that if it works with marbles, it will work with beans, fingers, toes, and so
on. We might make an inductive (experiment) to see whether on five runs of
counting a particular subject was always able to complete the task. We might
make predictions about human behavior (even our own) based on that. But with
the paradigm of calculation something else happens: we erect a way of looking
at things that draws something new into the realm of unthinkability: that there
could be 100 marbles (fingers, toes, etc.) and no possibility of a procedure for
counting 10 groups of 10. Here by "possibility" we do not mean a "medical"
or "experimental," but an atemporal conceptual or mathematical one.

The shift from experiment to calculation depends upon our ability to project such unthinkability, our ability to make it *real*. Wittgenstein has now
transformed his Uniformity Rule into a general investigation of human forms
of life.

The site of unformalized, everyday speech in mathematics, as elsewhere, is
precisely what traditional philosophy is driven to deny and question. While no
refutation of this denial is possible, and while the grammar of everyday speech
in mathematics is neither unrevisable nor unproblematic nor complete, it is
possible to ask us to attend in more fine-grained ways to what we actually,
ordinarily, do in mathematics, and also to attend to how it is that traditional
philosophical demands undercut this fine-grained attention. The deepest question about Wittgenstein's mature philosophy of mathematics is whether this
attention to what we do carries genuine philosophical significance, genuine
"reality."

4.3 Diagonal Proofs (RFM II)

In RFM II Wittgenstein explores Cantor's Diagonal Method as a *technique* of
proof. He investigates the extent to which it necessarily shows us something
general, for example, that a number of a certain kind is *not* contained in a
particular listing of numbers.[73] The issue is how far diagonalization allows us to
"transcend the limits" of a system. Wittgenstein's answer, influenced by Turing

---

[73] Commentary in WH Chapters 7–8.

(1936), is that it does not always do so, but that even when it does not, it may reveal a new aspect of a collection. We have already interpolated Wittgenstein's non-extensional emphasis on the variety of "samenesses" at work in particular diagonalizations regarded as formal series (§2.6, Figures 5–7). Now we revisit the Uniformity Principle from the Later Philosophy's point of view.

RFM II §2 considers the task, "Name a number that differs with $\sqrt{2}$ at every second decimal place." What kind of command *is* this? Is it obeyed if one says, "Develop $\sqrt{2}$ and add 1 or -1 to every second decimal place"? Yes, verbally, but it seems there is no aspect shift here taking us "beyond" where we were. Recall Middle Wittgenstein's "$\frac{5 \to 3}{\sqrt{2}}$", the rule: calculate the decimal expansion resulting from substituting "5" for "3" whenever it appears in the decimal expansion of $\sqrt{2}$ (§3.3). He questioned whether such a "purely formal" substitution gives us a new kind of number, suggesting that the "living trunk" of $\sqrt{2}$ is the vital core, given its connections with measurement and trigonometry (PR §182).

At issue with Cantor's proof is whether the diagonal method *per se* forces us to shift from a non-extensional to an extensional point of view. Wittgenstein rightly thinks it does not. Told to "divide an angle into three equal parts," we might simply have laid three equal angles together: there is no necessity in embedding the system of construction with ruler and compass in higher algebra and, having subjected it to analysis in these terms, declaring what we just did "unthinkable." We may take this step, and then the aspect shifts: it is now unthinkable to trisect the angle. But there is no way to *force* this projection from a new point of view (RFM II §2).

We might call this the "stubbornness" of the non-extensional perspective. It should be placed beside the stubbornness of the extensional perspective. There is no taking up both perspectives at once. So on one or another occasion we will have to choose our "form of life." Wittgenstein connects this point about occasion sensitivity to the "stubbornness" of humanity's differences amongst individuals and the stubbornness of our relying on known and familiar techniques – both also gifts to "realism" in mathematics, as we have seen (§4.2). Of course, Wittgenstein has his sympathies: he resists the "style" of extensionalism.[74] But his philosophical points are independent of this preference.

The diagonal method of Cantor is a *technique* of proof. From the extensional point of view, it shows that the cardinality of the real numbers is not countable, that there does not *exist* a 1-1 correlation with the natural numbers. Cantor will express the idea that the altered diagonal is "different from" all the previously

---

[74] Bangu, 2020.

listed extensions. For Wittgenstein this shifts the aspect under which we regard the listing of decimal expansions, but only if we go on to learn the calculus of infinite numbers Cantor fashioned, accepting 1-1 correlation as a criterion of numerical identity, and so on.

There is no problem with Cantor's proof as a proof. The problem is that, by a "skew form of expression," the extensionalist insists on seeing the result of the proof as our taking up a more extensionally adequate stance. That is fine from within her perspective. But the method of proof does not *establish* the perspective. The "skew" way of thinking puffs up Cantor's proof (RFM II §22).

What Cantor shows from the non-extensional point of view is that there is not just one space of techniques for expanding real decimal expansions ("I cannot teach you all techniques through a technique" (162b, p. 1). If we suppose we have listed all the expansions (indicated in a list all the rules), Cantor shows us a way of constructing a further rule not on the list: unbounded ways of expanding our arsenal of techniques for calculating out "new" decimal expansion representations of real numbers.

Thus the gloss that Cantor proved the uncountability of the real numbers is, so far as the diagonal argument technique goes, neither a logical must nor a mere "stipulation," as Putnam saddles Wittgenstein with saying.[75] The better angels of plasticity, the "faces of necessity" urged by Putnam himself, are *apropos*. Before Cantor's diagonal argument, simply being told that the real numbers are "uncountable" in the extensional sense ("there is no bijection from this set to the natural numbers") would have been mathematically empty. After Cantor has shown us the diagonal method and built his theory of cardinality, what once seemed empty may be regarded as necessarily true.

Wittgenstein's point is simply that Cantor's is not the only way to regard the situation. And he is correct: the Cantor argument may be regarded as a technique for constructing new decimal expansions. In fact, Hobson (1927, 90f.) regarded his non-extensional way of rendering the Cantor argument as its "completion": he takes Cantor to show that "the assumption of a *finite and complete* stock of words by which any element of the continuum may be finitely defined leads to contradiction."

Wittgenstein explored the diagonal technique in 1938–9 probably as a result of reading Turing's (1936) diagonal argument, which Turing sent to him by February 1937. Turing refers to Hobson in his §8, and Wittgenstein explicitly framed diagonalization as a recursively specifiable technique in a notebook (157a 17v, 1934–7).

---

[75] Putnam, 2012, 446.

Turing demonstrates that there can be no D machine, that is, a machine that would determine, Yes or No, for any Turing Machine T_i, whether or not the machine calculates an infinite expansion on input j.[76] His argument is not the familiar Halting argument, which derives a contradiction by building a "Contrary" machine that changes the coded outcome ("Halts" to "Doesn't Halt" and *vice versa*) of T_i on input j along the diagonal T_{ii}. Instead, Turing uses a *positive* diagonal argument in which the difficulty emerges directly from the characterization of the diagonal digits. He constructs a circular machine: when it gets to its own number, it cannot *do* anything.

Wittgenstein writes the proof down in terms of language-games:[77]

> A variant of [C]antor's diagonal proof:
> Let $v = \phi(k, n)$ be the form of the laws for the expansion of decimal fractions. $\underline{v}$ is the nth decimal place of the $\underline{k}$th expansion. The law of the diagonal then is:
> $v = \phi(n, n) =_{def.} \phi'(n)$.
>   It is to be proven that $\phi'(n)$ cannot be one of the rules $\phi(k, n)$. Assume it is the 100th. Then we have the formation rule
> of $\phi'(1)$: $\phi(1, 1)$
> of $\phi'(2)$: $\phi(2, 2)$
> etc.,
> but the rule for the formation of the 100th place of $\phi'(n)$ is/becomes $\phi(100, 100)$, that is, it tells us only that the 100th place is supposed to be equal to itself, and so for $n = 100$ is *not* a rule.

Wittgenstein thinks, non-extensionally, of "laws" or rules for decimal expansions placed on a list. $\phi(k, n)$ is the nth decimal place determined by the kth rule of expansion. For $n = 1$ the diagonal command $\phi'(n)$ says to calculate the first decimal place provided by the law $\phi(1, \ldots)$; for $n = 2$ to calculate the second decimal place provided by the law $\phi(2, \ldots)$, and so on. The question is whether any of the rules of the list correspond to $\phi'(n)$.

If we think extensionally, this argument goes nowhere. For all we know, there might be double counting going on in such a way that $\phi'(n)$ spells out an extension already on the list of "finished" expansions.

But if we imagine – as Turing and Wittgenstein do – that we are dealing with commands given to "machines" (i.e., human "computors") – then there is a problem. Suppose that $\phi'(n)$ already occurs somewhere on the list before we characterize the diagonal (e.g., as the 100th rule). Now the command for $n = 100$ is "*not* a rule" in the sense of the other rules on the list, for it cannot

---

[76] Floyd, 2012b, WH Chapter 7 for rigorous reconstructions.
[77] RPP I §1097, MS 135, 60ff.

be *followed*: it says "Do for $n = 100$ whatever $\phi'(100)$ tells you to do." But "Do What You Do" is empty as a command, in this context; it is reminiscent of "autological is autological," or the set of all sets that *are* members of themselves.[78] Therefore one cannot diagonalize *out* of the listing of computable real expansions.

Wittgenstein's diagonal argument shows that the living context in which a rule is applied must be taken into account for it to serve as a command: the embedding in a form of life helps us to *see* whether or not it *is* a rule that can be followed. "Computing a real decimal expansion" is an occasion-sensitive notion. Thus even though Turing's argument does *not* take us "out" of the system of computable real numbers, it shifts the aspect under which we regard calculating decimal expansions. The lesson is that a command cannot be shared without being associated with some kind of usable technique. We need a non-tautological occasion in order for tautologies to be used. There is no one technique for all techniques of expansion.

Wittgenstein regarded the self-reflexive aspect of Gödel's (1931) sentence analogously.[79] "This sentence is no tautology [cannot be derived in Russell's system] and it cannot be false" shifts the aspect under which we regard *Principia as* a system in relation to arithmetical truth: we "see the face of necessity" in his underivable sentence, which takes us beyond the idea of truth as "what is derivable in Russell's system" (121, 78vff). Gödel "confronts us with a new situation" (121, 84r).

Just before rehearsing the above-quoted version of Turing's (1936) argument, Wittgenstein reminds himself that "Turing's 'Machines'. These are *humans* who calculate." The human context is crucial for Turing's analysis of calculation-in-a-logic: it is *we* who bring the dynamism of a process or step-by-step routine *to* the Turing Machine. Regarded extensionally, a Turing Machine is simply a formalism or set of quadruples. Regarded non-extensionally, it is a model of a human computor calculating in accordance with a rule.

Wittgenstein takes Cantor's, Turing's, and Gödel's diagonal arguments to show the limits, not only of an extensional perspective on mathematics, but also on the very idea of a "purely mechanical" mode of thought. The point of Turing's (1936) is not to insist that we *are* machines, that the mind-machine "determines mechanically" all the steps that it can possibly take with its concepts: just the opposite. He explicitly *eliminates* any appeal to the notion of a "state of mind" by imagining written instructions on a piece of paper (§9 III), noting that "it is always possible for the computer to break off from his work,"

---

[78] Floyd, 2012b.
[79] Floyd, 2001.

leaving "a note of instructions (written in some standard form) explaining how the work is to be continued" by someone else.

Turing's notion of a "machine" is built for the Hilbertian context. It is a logico-mathematical one, not a psychological one. From Wittgenstein's point of view, Turing demonstrates that the human context of our capacity for colloquial language, a social art, is essential to calculation-in-a-logic (§3.1). Even if we can construct multiple computational routines to check one another's outcomes, we face at some point a basic circularity with the ideas, a need to acquiesce: the notion of *effectiveness* is not self-standing without colloquial elements in play.

The point is mathematically rigorous. Every Turing machine encoding a function does so with a particular scheme of notation (strokes, digits, placement of figures in electronic memory, etc.). Shapiro (1982) argues, while assuming a Turing idea of "effectiveness," that it is not possible to devise a decision procedure for distinguishing non-computable ("deviant") encodings from computable ones. Differently put, there is always a circularity in assuming that the relation between an integer and the expression coding it is "effective."[80] The notion of "performing a computation" is therefore occasion-sensitive.[81]

4.4 Surveyability (RFM III)

Commentators have long emphasized Wittgenstein's arguments against the idea that *Principia Mathematica*'s system *reduces* arithmetic to logic in the sense of explicating or justifying it.[82] His objection is that the proofs in *Principia* are "unsurveyable," whereas by contrast, as he repeatedly remarks, "a proof [or calculation] must be surveyable." The question is, what is the target, the content, and the force of this objection?

The issue is not whether knowing the axioms and definitions in *Principia* suffices to know arithmetic, or whether knowledge of arithmetic suffices to actually "grasp" all proofs in *Principia*. Wittgenstein is not making an a priori demand on proofs or knowledge or privileging elementary arithmetic in a brute, naturalistic way. He is not insisting on anything in particular *in* the order of knowledge. Rather, he focuses on shifts in our projections of possibilities and necessities for the concept of proof in mathematics: the dialectic of our techniques, our techniques with techniques, and so on. *Principia* brings

---

[80] Copeland and Proudfoot, 2010; Kennedy, 2017; Quinon, 2020.
[81] Turing, 1937, points out that there is no generally effective way of associating "description numbers" with real numbers; compare WH 230 for Wittgenstein's possible allusion to this.
[82] Marion, 2011, 2007; Mühlhölzer, 2006, 2010.

out a larger lesson and provides him with an opportunity to put his Later
views to work.

Most agree that Wittgenstein is not denying that basic arithmetical notions
and axioms may be formally defined in terms of "logic" in *Principia*, and
many theorems containing translations of ordinary arithmetical truths may be
obtained by logic alone in Russell's sense.[83] Arithmetic may be, in Wittgenstein's tendentious way of putting it, wrapped up in the "prose" of quantifiers,
predicate letters, and so on, and "tautologies" (pure logical deductions in the
system) thus constructed (RFM V §46). By 1939 Wittgenstein was aware that
not all truths may be so deduced, formally speaking (§4.6). But his focus on
Principia is on the non-extensional aspect of our notion of *proof*. How is this
notion actually used, embedded in mathematical forms of life, connected to
meaning and understanding?

Russell (1920, chap. 1) states that in logic alone may the determinacy and
applicability of arithmetical thoughts be secured: *Principia*'s individuation of
arithmetical content is the logicist's way of defying Dedekind's structuralism. RFM I, by contrast, recasts the idea of "determining the meaning" of
an expression as an occasion-sensitive, parochial notion demanding purposive techniques of fitting structured conceptual paradigms to forms of life. RFM
III applies this point of view to Russell's logicist "foundation."

For a human to calculate even an elementary sentence of arithmetic in *Principia* would take thousands of symbols. Like an unwieldy code, the expression
of thoughts explodes. The same is true if we try to rewrite the usual symbol-
izations of truth-functions in terms of the Sheffer Stroke, treating it as a mere
abbreviation (§3.3), or if we actually tried to use the stroke notation suggested
by Hilbert as a kind of intuitive basis for arithmetic (even "simple" expressions
become quickly unusable with pen and paper).

Hilbert sought "certainty" and intersubjective validity, hoping to control
mathematical controversy with the demand of a touchstone in the finite, concrete, given symbols of a basic notation. But his point was really that it is
mathematical techniques we require to distinguish, for example, ||||||||||||||||
from |||||||||||||||| or nonconstructive proofs from constructive ones. Wittgen-
stein accepts this point, while rejecting Hilbert's colorful rhetoric about "concrete" symbolic Gestalten giving us "grounding" for certainty, or articulating
"the" technique of thought as such.

The fact that arithmetical proofs may *possibly* be transcribed into *Principia*'s
(or *Principia* with Sheffer's) notation is logically and mathematically interest-
ing, pointing toward the mechanization of proof. But that interest is not to be

---

[83] But see Steiner, 1975.

articulated by actually writing out formal proofs day after day, as if *that* secures certainty, increases the inductive confirmation of the transposition, or unpacks needless abbreviations (§3.3).

Unwieldy formal proofs are not used to check elementary arithmetical sums; rather we index quantifiers so that we apply arithmetic *to* a *Principia* proof, for example, by counting parentheses (If $(\exists_{17}(x)$ and $(\exists_{24}(x))$ then $(\exists_{41}(x))$. We prefer our "everyday" notation to reveal aspects of *Principia* notation, because we require that what the formal proof produces may be communicated without squabbling or questioning. If our purposes involve vast numbers, we find our way with different techniques. We work in colloquial language to draw out the interest of a proof (§3.1).

Wittgenstein does not discuss computer-assisted proofs (though see RFM IV §20). Yet his criticisms of overly simplified conceptions of understanding fit challenges that arise nowadays in higher-level languages for proof verification as well as interactive programming.[84] The dialectical interplay in demands for simplicity and surveyability here confirms the "multicolored" character of our techniques for "understanding the meaning" of a concept or proof (RFM III §46ff.). The distinction between "calculation" and "experiment" shades off in computer proofs. The unabbreviated level of machine language serves implementation ends that, coupled with domain-specific understandings and human–machine interaction, allow us to assemble and reassemble concepts, furthering our own mathematical understandings.

Today the need to utilize mechanically generated models and pictures to take in, generate, and communicate vast quantities of data is an everyday phenomenon in our implementation of language, one that serves pedagogical, social, and political ends. These may be brought into play to "determine" meanings and secure "understandings." As computer science arguably becomes a branch of social science, Wittgenstein's turn toward "surveyability" becomes ever more relevant to "foundations."

But what is the philosophical import of this dynamic, modal understanding of "surveyability"? By a proof (or calculation) being "surveyable" Wittgenstein construes it as a *picture* of what *can* (or cannot) be done. The modality expresses his idea that the "limits" of empiricism lie in the imperative to communicate, in concept-formation (RFM III §71, VI §23), and not in the "medical impossibility" (e.g.) of our surveying or computing an infinite number of calculations in finite time (Russell, 1936). Wittgenstein's aspect-talk provides a way to speak of "experience" or "intuition" in the epistemology of mathematics both in terms of immediate apprehension of a particular and in terms of a

---

[84] Shanker, 1987; Avigad, 2008a, 2008b.

procedure unfolding: it focuses on action, what can (and cannot) be done, ways in which we compare, discuss, characterize, and regard things from a variety of angles (RFM VII §21).

Turing remarked that his arguments rely *directly* on "intuition" and "experience," something "not ideal" mathematically. He assumes, for example, that his computors must be able to "take in at a glance" a meaningful configuration of symbols [§9]OCN. Wittgenstein is not against this way of speaking – in fact, he unpacks the "surveyability" of proof with the idea of a proof being "intuitive," or an "intuitive" procedure – but by "intuitive" Wittgenstein means something "homespun," down to earth, like "plain to view" (RFM III §42, IV §30).

It is "not something behind the proof," but the proof itself, with its specific configuration of expressions, that proves: the proof can be taken in as we use it to project necessity and possibility. A proof may "remodel" our intuition, in the sense of getting us to a new *way* of looking at things (RFM IV §30). Therefore, a proof is not an "interpretation" of signs but a projection of a dimension of conceptual configuration: it changes the aspects of concepts, fashioning unthinkability and unimaginability (RFM III §41). Wittgenstein's interest is in unpacking the complexities and practical facets of the indirection involved in such projection. He does not think "surveyability" is always easily achieved.

Thus he uses an analogy between procedures of proof and the dynamism of film projection to bring out the complex ways in which the necessities *in* procedures (rather than temporal movements) may be exemplified. It is part of the concept of "proof" that we aim at clarity in the unfolding of a possible procedure that creates a home for a concept: the steps must be able to be communicated, reproduced without controversy.

In films that teach (say YouTube How-To Videos), the goal and the basic tools are first assembled; uninteresting or lengthy temporal steps in a procedure are cut or speeded up on film as of negligible interest *to* the procedure; focus, movement of the camera angle, and images of humans performing the relevant actions help convey a narrative for the step-by-step routine with emphases, things to avoid, and so on. If the issue becomes whether the film's narrator "fully understands" the proof, this can come out in a variety of different ways: objections and questions may be answered, examples provided, an outline of the idea of the proof articulated, and so on. The film erects criteria of identity for "proceeding in the same way."

Wittgenstein does not mean that every proof can be "taken in at a glance" in the sense of easiness, memorizability, or visual simplicity. It is not always clear whether a proof *can* be re-rendered to become easier to take in or more feasible to use. Many mathematical arguments are difficult to take in and do not contain pictures or diagrams in the literal sense at all. What is difficult for imagination at first becomes more insightful with further knowledge (and computational power). Nevertheless, communication is a primary desideratum, the bringing of mathematical proof as far as we can toward an ideal where we can lean on everyday, colloquial language, with all its dynamism and potential vagueness. Turing (1942–4) reported that he learned this very point from Wittgenstein's lectures.[85]

4.5 Real Numbers (RFM V)

Because of the occasion-sensitivity, Later Wittgenstein emphasizes as fundamental to logic and mathematics, many readers believe that he subscribes to verificationism in spite of himself: the meaning of a proposition *is* its proof. It is a quick step from here to the "anti-realism" of Dummett and Wright.[86]

The trouble with this reading is that it underplays the multidimensional sense of "occasion" Wittgenstein has in mind, his aspectual realism, which emphasizes the multiple analyzability of mathematical concepts, including the "tensions" involved in balancing our presentation of the real numbers (§3.5). A proof may reveal one *side* of a number or structure but leave us in the dark about another. Insight and understanding in mathematics is neither flat, nor given by sentences alone, but given through multidimensional aspects that are brought out in techniques of their projection. If understanding is only always partial, that does not mean that it is not real.

Middle Wittgenstein *was* a verificationist, and in a fairly clear sense an "anti-realist" (§3.3). Later Wittgenstein advocates a more fluid approach to sense-individuation, regarding the criteria for samenesses and differences among proofs of "the same proposition" as something that we can settle in discussion, on different occasions, determining as we go whether or not a proof simply recapitulates the idea of another proof, or whether it "does something new."

It is an everyday mathematical fact that there can be debate about what counts as "the same proof" and that progress may be made in discussion of particular cases, and that sometimes final agreement may not be reached. This *confirms* Wittgenstein's idea of the fluidity of simplicity. A relevant example is Turing (1936): although Church (1936) had already resolved the *Entscheidungsproblem* in the negative by the time Turing showed his proof to Newman, Turing's *way* of resolving "the same thing" used such a different approach that it was still published. Turing had to add an Appendix to his paper proving that his "Turing Machines" can compute any function that is λ-computable in Church's sense,

---

[85] Floyd, 2013.
[86] Wright, 1980, 2001.

extensionally speaking. But he was still allowed to publish – the aspectual point (§4.6).

We have already seen Wittgenstein differentiating between ways of regarding Cantor's diagonal procedure with infinite expansions and the diagonal technique itself. Different or "the same" things are proved depending upon which perspective we take up. When we see the other side of a concept or proposition, we project differently, but we do not always say that there is a completely different technique of proof at stake. Conversely, though there may be different techniques of proof, we can sometimes say that the result reached is "the same." An example is again Cantor's diagonal argument. It is very different on the surface from Cantor's original (1874) topological way of arguing for the uncountability of the real numbers, not at all "the same proof." But retrospectively, with his (1891) in hand, we can see in the collapse of two indices into one an implicit "diagonalization" (WH 133n, 174, 214).

Did Later Wittgenstein fall back into verificationism? According to Putnam, Yes. He cites (RFM V §41):

> What harm is done e.g. by saying that God knows all irrational numbers? Or: that they are already all there, even though we only know certain of them? Why are these pictures not harmless?
>
> For one thing, they hide certain problems. —
>
> Suppose that people go on and on calculating the expansion of π. So God, who knows everything, knows whether they will have reached "777" by the end of the world. But can his omniscience decide whether they would have reached it after the end of the world? It cannot. I want to say: Even God can decide something mathematical only by mathematics. Even for him the mere rule of expansion cannot decide anything that it does not decide for us.
>
> We might put it like this: if the rule for the expansion has been given us, a calculation can tell us that there is a '2' at the fifth place. Could God have known this, without the calculation, purely from the rule of expansion? I want to say: No.

Update the remark: computers have already found an occurrence of "777" in the decimal expansion of π. Note that Wittgenstein is tentative, remarking that he "wants to say No" – but does not say it.

Putnam insists that Wittgenstein committed himself in this remark (perhaps unwittingly) to the view that humanly undecidable mathematical propositions lack a truth value, a "deeply problematic" idea tantamount to embracing a verificationist account of mathematical truth and thereby denying realism (2012, 356, 432). Some true differential equations of physics are likely impossible to solve by calculation or are uncomputable, yet we say their solutions exist, either do or do not hold of physical reality (2012, 221f., 375, 431n.). For Putnam, "to say that omniscience cannot decide a question is ...Wittgenstein's way of saying that there is no answer to be known, i.e., it is an illusion that there must be a right answer" (2012, 356n., 431f.).

But Wittgenstein does not say this. Instead he warns of the "harm" certain pictures can do *if* they "hide certain problems." The problems he has in mind involve the confusion of the non-extensional and the extensional perspectives. The idea that "God knows all irrational numbers" sounds all right, but the "all" and the kind of acquaintance involved by its means are ambiguous.

There are several problems with the thought that from the decimal rule of expansion of the digits of π God would know "all" particular digits of π that humans would have ideally calculated by the end of the world – thereby somehow similarly knowing "all" irrational numbers. One is that not all irrational numbers may be represented in decimal expansion form if we restrict ourselves to expansions calculable by a rule.

A deeper problem is the mixing up of differing models, differing aspects, of the real numbers. The interlocutor couples God both with the extensionalist idea of "seeing at a glance" *all* digits of π and with the idea of God as an omniscient physicist/anthropologist. But this conflates logic, causality, and the idea of human techniques: even God should be careful to separate these ways of thinking. Cantor may be taken to have shown us something about human techniques and the languages we employ to articulate our understandings of the real numbers – i.e., there is no general technique that can show us *all* the techniques of expansion we might concoct (§4.3). Moreover, there remains the "tension" of which Bernays wrote (§3.5): the need to balance an understanding of the continuum as a whole with the uses and construals of real numbers as individuals.

We have already seen how unclear Wittgenstein finds the idea of "human calculating the decimal expansion of π": there could be a variety of substitutive routines grafted onto the usual ways of, for example, calculating a particular irrational decimal expansion, and it is unclear which would be "the same," and which "different' (§3.3, §4.3).

Wittgenstein's point is that the richness of mathematical aspects – the differing faces that the real numbers show us – should not be "harmed," that is, suppressed or flattened out, by too-quick ways of thinking of understanding, meaning, and calculation and fitting together in one way. The occasion-sensitivity and plasticity of our concepts is what drives our impulse to characterize and recharacterize in mathematics in the first place, to fashion new techniques. The interlocutor wipes out this richness and supposes that the extensional point of view will make it inessential. In so doing, she loses what for Wittgenstein are the "realities" of the real numbers.

Wittgenstein's immediately following remark (RFM V §42) conveys something quite different from the verificationism Putnam reads into him:

> When I said that the propositions of mathematics determine [*bilden*] concepts, that is vague; for "2 + 2 = 4" forms a concept in a different sense from "$p \supset p$", "$(\forall).fx \supset fa$", or Dedekind's Theorem. The point is, there is a family of cases. (RFM V §42)

Later Wittgenstein's point is not that incalculable decimal expansions are "indeterminate" or that our calculations "lack a truth-value." Rather, mathematical concepts are variously structured and truth-values of statements must respect this. If God has to decide mathematical questions through mathematics, then even God must respect the fact that there is other mathematics that is appropriate to use here. The idea of a rule of expansion for digits of π does not tell us, in the relevant way, what "all" digits of π are, much less does the notion of a rule of decimal expansion tell us "in general" what "all" irrational numbers are (RFM V §9, VI 41; WH 3.8). Our mathematical practices consist of a multicolored collection of techniques.

This is why Wittgenstein adduces Dedekind's theorem, which construes the totality of real numbers as classes of rationals. He is not rejecting Dedekind's notion of a cut, even if he finds the extensional presentation of it misleading (RFM V, WH §3.1). He is not restricting the meaningful concepts of mathematics to physically calculable ones. Rather, he resists a "harmful" and misleading mixture of pictures. The extensional point of view, used unreflectively, encourages a disrespect for the plasticity of our concepts.

What is merely an "illustration," and what a true "application" of a concept? The question arises at the outset of Wittgenstein's 1942–3 annotations to Hardy (1941) (WH §§3.1–2). Hardy states (p. 2) that the geometrical line is merely an "illustration," with no systematic significance for analysis. Wittgenstein counters that while "illustration" is inessential, "application" is not (WH §5.5). He does not mean by this "application in physics," as Putnam thinks. What he means is that geometrical considerations have ceased to function in anything like a straightforward way in modern analysis since Dedekind and Cantor: analysis is severed from the notion of quantity.

Hardy (1941, 8) dismisses the possibility of introducing decimal expansions as being "hardly of the precise character required by modern mathematics." But for Wittgenstein decimal expansions are standard representations of irrational numbers. Even from a really rigorous point of view, decimal expansions may be used to represent a perfectly respectable transition from the rational to all the real numbers.[87]

The (extensionalist) idea of decimal expansions of real numbers as "finished" nicely fits the geometrical image of the straight line: the finished expansions represent points on the line. To consider only the rational numbers leaves open gaps, and from this point of view the filling of these gaps with irrational numbers appears to be "prejudged" (§3.5). Yet this "gap" imagery can be misleading in being circular. For the extensional point of view *can only* regard the shift from considering the rationals to the reals *as* a widening of the extension of the concept "real number"; it cannot regard the shift in any other way. *That* is what Wittgenstein is pointing out.

4.6 Wittgenstein, Gödel, and Turing (RFM I, II, V, VII)

Later Wittgenstein responded guardedly and haltingly to the incompleteness and undecidability results of Gödel (1931) and Turing (1936).[88] His understanding of these results, however partial, helped him shift to his Later philosophy. Now he regards mathematics, as well as philosophy, as "a class of *investigations*," that is, explorations where the clarity and/or extent of our concepts is not taken for granted as clear in advance, but articulated and rearticulated indefinitely (117, 51ff.).[89] At the same time he continues to work out his idea of occasion sensitivity, the multiple-aspect view of sentences, applying many different arguments to our uses of sensation concepts, perceptions, and even first-person ascriptions of experience.

Wittgenstein remarks that Gödel has "done a great service to philosophy," shifting the aspect under which we view mathematics (163, 39v-40v). The need for reformulations of what is meant by "is a proof" in a variety of contexts is shown in Gödel's (1931), especially in the second incompleteness theorem, where characterization of "consistency" and "provable" becomes an actual task.[90] Middle Wittgenstein's relativism, by contrast, made the consistency of the elementary propositions *within* a system necessary.

Often Wittgenstein is regarded as quarreling with Gödel, but that is because he is too often read as a radical finitist or conventionalist. His remarks certainly struggle to place Gödelian incompleteness into his way of thinking. But if we take seriously his Later views, we see what the struggle is about, and it is not

---

[87] Gowers, 2017b; Tao, 2006, 380–388.
[88] Rodych, 2002 surveys the manuscript remarks on Gödel.
[89] Floyd, 2001; Floyd and Putnam, 2000.
[90] WCL 50ff.; Floyd and Putnam, 2000, 2012.

about refuting Gödel. There is a richness to the ways we may articulate arithmetic, as Gödel proved, and a richness to the idea of "follows from a set of axioms in a formal language," as Church and Turing showed. Conventionalists, formalists, and logicists should admit and learn from this. For the mature Wittgenstein, the multidimensional play comes out in our articulations of mathematics in everyday phraseology and its embedding in life, in the techniques we establish and share. This is not an alternative to formalizing theories where we can, but it is not wholly reducible to that activity.

Later Wittgenstein sees both a pluralism and a universalism in our notion of logic. The price of unity is the fluidity of simplicity. Logic has not the formal unity he assumed it did in TLP; nor is it, however, an archipelago of independent calculi. Instead, it evolves and is a "multicolored" collection of techniques for projecting, building, structuring, and restructuring concepts.

Turing concocted an "object of comparison," a language-game to resolve the *Entscheidungsproblem*: he analogized a human being reckoning according to a fixed rule "mechanically" to a machine. This brought the whole idea of a formal system down to earth, making it *plain*. He could convincingly analyze, with a mimimum degree of formalization, what is *meant* by the idea of a "step" in a formal system of logic. Turing found the right parametrization of our concepts by thinking what (and who) calculation-in-a-logic is *for*.

Although Gödel, Church, Kleene, and Rosser clarified the notion of "effective" before Turing, they were mathematicians thinking in terms of systems of equations and the notation of functions. Turing's analysis alone, in bringing the human being's purposes into the center, is immediate and intuitive. This served a purpose. In order to analyze what a formal system of logic *is*, one could not simply write down another formal system. One had to *do* something: shift the *aspect* under which we regarded what the *Entscheidungsproblem* required. Turing's diagonal argument is free of any tie to a particular formalism or symbolism. And it depends upon an everyday idea about rules, rule-following, and sharing of commands, but *not* on any community-wide agreement or consensus.

Turing's diagonal argument (§4.3) shows that it is not an essential part of our notion of taking a correct step in a formal system of logic (or calculating according to a rule) that one does or does not subscribe to the law of excluded middle in infinite contexts. Even an intuitionist can accept his positive diagonal argument, the "Do What You Do" form of reasoning (§4.3). Nor is a theory of truth required to accept his analysis. Nor any particular symbolism. What matters is the analogy between a human being, reckoning step-by-step according to a fixed rule, and a machine. Turing's success is partly extensional: he had to prove that all of the Turing-computable functions are λ-computable, and vice versa. But the most profound aspect of his success is non-extensional: he refashions and re-projects our concepts vividly.

Gödel agreed. Not until Turing clarified what is *meant* by a "formal system" of the relevant kind was the extent of applicability of Gödel (1931) clear. As Gödel remarked, because of Turing's

> precise and unquestionably adequate definition of the general concept of formal system, ...the existence of undecidable arithmetical propositions and the non-demonstrability of the consistency of a system in the same system can now be proved rigorously for every consistent formal system containing a certain amount of finitary number theory.[91]

Equally crucial, because one cannot diagonalize out of the class of Turing-computable functions, the notion of "computable" is *absolute*: it does not vary with the formal system. With Turing's analysis,

> one has for the first time succeeded in giving an absolute definition of an interesting epistemological notion, i.e., one not depending on the formalism chosen... In all other cases treated previously, such as demonstrability or definability, one has been able to define them only relative to a given language, and for each individual language it is clear that the one thus obtained is not the one looked for. For the concept of computability, however, although it is merely a special kind of demonstrability or decidability, the situation is different. By a kind of miracle it is not necessary to distinguish orders, and the diagonal procedure does not lead outside the defined notion (Gödel, 1946, 150).

But it is hardly a "miracle" that Turing's Wittgensteinian, down-to-earth approach defeats transcendence through diagonalization. Turing took up a form of life perspective on the idea of a formalism. The unity of logic is a task, constantly under repair, as we proceed through life. And "states of mind" are shareable commands that may be *followed*.[92]

Turing's Universal Machine does the work of all machines, but it cannot be diagonalized out of. This indicates the ubiquity and indefinite applicability of computational processes and modeling in our world. There are no sharp dichotomies to be drawn between software, hardware, and data: the Universal Machine can "adapt" to its own outputs, inputs, and internal commands without limit, in the manner of our computers and certain biological, physical, and social processes today.[93]

---

[91] Gödel, 1964b.
[92] Floyd, 2017.
[93] Davis, 2017.

Turing said he had learned from Wittgenstein's lectures about the importance of everyday language, that is, evolving phraseology, as a factor in the evolution of logical "types."[94] The point connects directly with the "realism" we have ascribed to Wittgenstein. Turing's "computors" must be able to "take in at a glance" a configuration of symbols. But the indefinite range of symbolic compression for us, given computers and our social networks, makes this an evolving target. We must surrender the ideal of a universal logical language of the kind the TLP demanded. Instead, we should regard ourselves embarked on what Turing (1948) described as "the cultural search": the drive for new techniques, creativity in phraseology, and "initiative," (i.e. new forms of life).

---

[94] Turing, 1942–4; Floyd, 2013.

Abbreviations

BB, BrB, BBB *The Blue and Brown Books*
BEE, MS, TS Bergen Edition of Wittgenstein's *Nachlass*
LFM *Wittgenstein's Lectures on the Foundations of Mathematics, Cambridge 1939*
MWL *Wittgenstein: Lectures, Cambridge 1930–1933*
NB *Notebooks 1914–1916*
PG *Philosophical Grammar*
PI, PPF *Philosophical Investigations*
PR *Philosophical Remarks*
RFM *Remarks on the Foundations of Mathematics*
RLF *Some Remarks on Logical Form*
RPP I *Remarks on the Philosophy of Psychology I*
TLP *Tractatus Logico-Philosophicus*
WH *Wittgenstein's Annotations to Hardy's Course of Pure Mathematics*: Floyd and Mühlhölzer, 2020
WVC *Wittgenstein and the Vienna Circle*

References

Auxier, R. E., Anderson, D. R., & Hahn, L. E. 2015. *The Philosophy of Hilary Putnam*. Chicago, IL: Open Court.

Avigad, J. 2008a. *Computers in Mathematics*. (In Mancosu, 2008, pp. 300–316.)

Avigad, J. 2008b. *Understanding Proofs*. (In Mancosu, 2008, pp. 317–353.)

Bangu, S. 2020. "Changing the Style of Thinking": Wittgenstein on Superlatives, Revisionism, and Cantorian Set Theory. *Iyyun, the Jerusalem Philosophical Quarterly, 68*, 339–359.

Barwise, J. (ed.). 1977. *Handbook of Mathematical Logic*. Amsterdam; New York: North-Holland Pub. Co.

Baz, A. 2020. *The Significance of Aspect Perception: Bringing the Phenomenal World into View* (vol. 5). Switzerland: Springer International Publishing AG, Springer.

Benacerraf, P. 1965. What Numbers Could Not Be. *The Philosophical Review, 74*(1), 47–73.

Bernays, P. 1957. Bemerkungen zum Paradoxon von Thoralf Skolem. *Avhandlinger utgitt av Det Norske Videnskaps-Akademi i Oslo, I. Mat.-Naturv. Klasse, 2nd series*, 3–9. (English translation by Dirk Schllimm and Steve Awodey, "Considerations regarding the paradox of Thoralf Skolem," www.phil.cmu.edu/projects/bernays/.)

Bernays, P. 1959. Betrachtungen zu Ludwig Wittgensteins *Bemerkungen über die Grundlagen der Mathematik*/Comments on Ludwig Wittgenstein's *Remarks on the Foundations of Mathematics*. *Ratio, 2*(1), 1–22.

Brouwer, L. E. J. 1929. Mathematik, Wissenschaft und Sprache. *Monatshefte für Mathematik und Physik, 36*(1), 153–164.

Cantor, G. 1874. Über eine Eigenschaft des Inbegriffes aller reellen algebraischen Zahlen. *Journal für die reine und angewandte Mathematik, 77*, 258–262.

Cantor, G. 1891. Über eine elementare Frage der Mannigfaltigkeitslehre. *Jahresbericht der Deutschen Mathematiker-Vereinigung, I*, 75–78. (English translation in Ewald, 1996, vol. 2, pp. 920–922.)

Cavell, S. 1988. Declining Decline: Wittgenstein as a Philosopher of Culture. *Inquiry, 31*(3), 253–264.

Church, A. 1932. A Set of Postulates for the Foundation of Logic. *Annals of Mathematics, 33*(2), 346–366.

Church, A. 1936. A Note on the *Entscheidungsproblem*. *Journal of Symbolic Logic, 1*(1), 40–41.

Copeland, B. J. 2002. Accelerating Turing Machines. *Minds and Machines, 12*(2), 281–300.

Copeland, B. J. & Proudfoot, D. 2010. Deviant Encodings and Turing's Analysis of Computability. *Studies in History and Philosophy of Science, 41*(3), 247–252.

Costreie, S. (ed.) 2016. *Early Analytic Philosophy: New Perspectives on the Tradition.* New York: Springer Publishing Switzerland.

Crocco, G., & Engelen, E.-M. (eds.) 2016. *Kurt Gödel: Philosopher-Scientist* (vol. 1). Aix-en-Provence: Presses Universitaires de Provence.

Davis, M. 2017. *Universality Is Ubiquitous.* (In Floyd and Bokulich, 2017, pp. 153–158.)

Dedekind, R. 1872. *Stetigkeit und irrationale Zahlen.* Wiesbaden: Friedrich Vieweg und Sohn. (English translation *Continuity and Irrational Numbers* in Ewald, 1996, vol. 2, pp. 766–779.)

Diamond, C. 1991. *The Realistic Spirit: Wittgenstein, Philosophy, and the Mind.* Cambridge, MA: Massachusetts Institute of Technology Press.

Dreben, B., & Floyd, J. 1991. Tautology: How Not to Use a Word. *Synthese, 87*(1), 23–50.

Dummett, M. A. E. 1959. Wittgenstein's Philosophy of Mathematics. *The Philosophical Review, LXVII*, 324–348.

Dybjer, P., Lindström, S., Palmgren, E., & Sundholm, G. (eds.) 2012. *Epistemology versus Ontology, Logic, Epistemology: Essays in Honor of Per Martin-Löf.* Dordrecht: Springer Science+Business Media.

Ebert, P. A., & Rossberg, M. (eds.) 2019. *Essays on Frege's Basic Laws of Arithmetic.* Oxford: Oxford University Press.

Ellis, J., & Guevara, D. (eds.) 2012. *Wittgenstein and the Philosophy of Mind.* New York/Oxford, UK: Oxford University Press.

Engelmann, M. L. 2013. *Wittgenstein's Philosophical Development: Phenomenology, Grammar, Method, and the Anthropological View.* Basingstoke, UK: Palgrave Macmillan.

Ewald, W. (ed.). 1996. *From Kant to Hilbert: A Source Book in the Foundations of Mathematics.* New York: Oxford University Press (2 vols.).

Feferman, S. 2005. *Predicativity.* (In Shapiro, 2005, pp. 590–624.)

Fisher, D., & McCarty, C. 2016. *Reconstructing a Logic from Tractatus: Wittgenstein's Variables and Formulae.* (In Costreie, 2016, pp. 301–324).

Floyd, J. 1995. *Wittgenstein, Gödel and the Trisection of the Angle.* (In Hintikka, 1995, pp. 373–426.)

Floyd, J. 2001. Prose versus Proof: Wittgenstein on Gödel, Tarski and Truth. *Philosophia Mathematica, 3*(9), 280–307.

Floyd, J. 2005. *Wittgenstein on Philosophy of Logic and Mathematics.* (In Shapiro, 2005, pp. 75–128.)

Floyd, J. 2010. *On Being Surprised: Wittgenstein on Aspect Perception, Logic and Mathematics.* (In Krebs and Day, 2010, pp. 314–337.)

Floyd, J. 2011. Prefatory Note to the Frege–Wittgenstein Correspondence. In E. De Pelligrin (ed.), *Interactive Wittgenstein* (pp. 1–14). Springer Science + Business Media B.V.

Floyd, J. 2012a. *Das Überraschende: Wittgenstein on the Surprising in Mathematics.* (In Ellis and Guevara, 2012, pp. 224–258.)

Floyd, J. 2012b. *Wittgenstein's Diagonal Argument: A Variation on Cantor and Turing.* (In Dybjer, Lindström, Palmgren, and Sundholm, 2012, pp. 25–44.)

Floyd, J. 2013. *Turing, Wittgenstein and Types: Philosophical Aspects of Turing's 'The Reform of Mathematical Notation' (1944–5).* (In Turing, 2013, pp. 250–253.)

Floyd, J. 2016. Chains of Life: Turing, *Lebensform*, and the Emergence of Wittgenstein's Later Style. *Nordic Wittgenstein Review, 5*(2), 7–89.

Floyd, J. 2017. *Turing on "Common Sense": Cambridge Resonances.* (In Floyd and Bokulich, 2017, pp. 103–152.)

Floyd, J. 2018a. *Aspects of Aspects.* (In Sluga and Stern, 2018, pp. 361–388.)

Floyd, J. 2018b. *Lebensformen: Living Logic.* (In Martin, 2018, pp. 59–92.)

Floyd, J. 2020. Aspects of the Real Numbers: Putnam, Wittgenstein, and Nonextensionalism. *The Monist, 103*(4), 427–441.

Floyd, J. 2021. *Sheffer, Lewis and the "Logocentric Predicament."* (To appear in J. P. Narboux and H. Wagner, eds., *The Legacy of C. I. Lewis*, Routledge Series in American Philosophy.)

Floyd, J., & Bokulich, A. (eds.) 2017. *Philosophical Explorations of the Legacy of Alan Turing – Turing 100.* Dordrecht: Springer.

Floyd, J., & Kanamori, A. 2016. *Gödel vis-à-vis Russell: Logic and Set Theory to Philosophy.* (In Crocco and Engelen, 2016, pp. 243–326.)

Floyd, J., & Mühlhölzer, F. 2020. *Wittgenstein's Annotations to Hardy's Course of Pure Mathematics, An Investigation of Wittgenstein's Non-Extensionalist Understanding of the Real Numbers.* Switzerland: Springer.

Floyd, J., & Putnam, H. 2000. A Note on Wittgenstein's 'Notorious Paragraph' about the Gödel Theorem. *Journal of Philosophy, 45*(11), 624–632.

Floyd, J., & Putnam, H. 2012. *Wittgenstein's "Notorious" Paragraph about the Gödel Theorem: Recent Discussions.* (In Putnam, 2012, pp. 458–481.)

Fogelin, R. J. 1987. *Wittgenstein* (2nd ed.). London; New York: Routledge K. Paul. (1st ed., 1976.)

Fogelin, R. J. 2009. *Taking Wittgenstein at His Word: A Textual Study* (vol. 29). Princeton, NJ: Princeton University Press.

Frascolla, P. 1997. The *Tractatus* System of Arithmetic. *Synthese, 112*, 353–378.

Frege, G. 1879. *Begriffsschrift, eine der arithmetischen nachgebildete Formelsprache des reinen Denkens*. Halle: Louis Nebert. (Reprinted 1964 by G. Olms, Hildesheim. English translation by Stefan Bauer-Mengelberg in van Heijenoort, 1967, pp. 5–82 and Preface and Part I by Michael Beaney in Frege, 1997, pp. 47–78.)

Frege, G. 1880/1881. *Booles rechnende Logik und die Begriffsschrift (1880/1881*. (In Frege [1969], pp. 9–52. English translation "Boole's Logical Calculus and the Concept-script" in Frege, 1979, pp. 47–52.)

Frege, G. 1884. *Die Grundlagen der Arithmetik: eine logisch-mathematische Untersuchung über der Begriff der Zahl*. Breslau: W. Koebner. (English translation by J. L. Austin, *The Foundations of Arithmetic: a Logico-Mathematical Enquiry into the Concept of Number*, New York: Harper Torchbooks, 2nd rev. ed., 1953.)

Frege, G. 1891. *Funktion und Begriff*. Jena: Hermann Pohle. (In Frege, 1969, pp. 125–143. English translation by Peter Geach in Frege, 1997, pp. 130–148.)

Frege, G. 1892. Über Sinn und Bedeutung. *Zeitschrift für Philosophie und philosophische Kritik, 100*, 25–50. (English translation by Max Black, "On *Sinn* and *Bedeutung*," in Frege, 1997, pp. 151–171.)

Frege, G. 1893/1903. *Grundgesetze der Arithmetik; Begriffsschriftlich Abgeleitet, Volumes I and II*. Jena: Verlag von Hermann Pohle. (Edited and translated into English as *Basic Laws of Arithmetic* by Philip A. Ebert, Marcus Rossberg, and Crispin Wright, Oxford University Press, Oxford, 2013.)

Frege, G. 1897. *Logik*. (In Frege, 1969, pp. 137–163. English translation by Peter Long and Roger White, "Logic (1897)," in Frege, 1997, pp. 227–250.)

Frege, G. 1918/19. Der Gedanke. *Beiträge zur Philosophie des deutschen Idealismus, I*, 58–77. (English translation by Peter Geach and R. H. Stoothoff in Frege, 1984, pp. 342–362 and in Frege, 1997, pp. 325–346.)

Frege, G. 1969. *Frege: Nachgelassene Schriften, Band I*. Hamburg: Felix Meiner Verlag. (Edited by Hans Hermes, Friedrich Kambartel, and Friedrich Kaulbach.)

Frege, G. 1979. *Posthumous Writings*. Chicago: University of Chicago Press.

Frege, G. 1980. *Philosophical and Mathematical Correspondence*. Oxford: B. Blackwell. (Edited by Gottfried Gabriel, Hans Hermes, Friedrich Kambartel, Christian Thiel, Albert Veraart; abridged from the German edition by Brian McGuinness, translated by Hans Kaal.)

Frege, G. 1983. *Wissenschaftlicher Briefwechsel* (Edited by Gottfried Gabriel, Hans Hermes, Friedrich Kambartel, and Christian Thiel). Hamburg: F. Meiner. (English translation in Frege, 1980.)

Frege, G. 1984. *Collected Papers on Mathematics, Logic, and Philosophy*. New York: Oxford University Press.

Frege, G. 1997. *The Frege Reader*. Malden, MA: Blackwell Publishers. (Edited by Michael Beaney.)

Gödel, K. 1930. Die Vollständigkeit der Axiome des logischen Funktionenkalküls. *Monatshefte für Mathematik und Physik, 37*, 349–360. (Reprinted with English translation in Gödel, 1986, 103–123.)

Gödel, K. 1931. Über formal unentscheidbare Sätze der *Principia Mathematica* und verwandter System I. *Monatshefte für Mathematik und Physik, 38*, 173–198. (Reprinted with English translation in Gödel, 1986, 144–195.)

Gödel, K. 1946. *Remarks Before the Princeton Bicentennial Conference on Problems in Mathematics*. (Reprinted with English translation in Gödel, 1990, 150–153.)

Gödel, K. 1964a. *On Undecideable Propositions of Formal Mathematical Systems*. (Lectures given at the Institute for Advanced Study, February–May 1934. Transcribed by S. C. Kleene and J. B. Rosser. In Gödel, 1986, pp. 346–369, with a Postscriptum, namely Gödel, 1964b.)

Gödel, K. 1964b. *Postscriptum to the 1934 Princeton Lectures, 1964*. (Postscriptum added June 3, 1964 to the reprinting of Gödel, 1964a, in Gödel, 1986, 369–371.)

Gödel, K. 1986. *Collected Works, Volume I: Publications 1929–1936*. Oxford: Clarendon Press. (Edited by Solomon Feferman *et al.*)

Gödel, K. 1990. *Collected Works, Volume II: Publications 1938–1974*. Oxford: Clarendon Press. (Edited by Solomon Feferman *et al.*)

Goldfarb, W. 2018a. *Moore's Notes and Wittgenstein's Philosophy of Mathematics: The Case of Mathematical Induction*. (In Stern, 2018, pp. 241–252.)

Goldfarb, W. 2018b. *Wittgenstein against Logicism*. (In Reck, 2013, pp. 171–184.)

Goodstein, R. 1957a. Critical Notice of *Remarks on the Foundations of Mathematics*. *Mind, LXVI*, 271–286.

Goodstein, R. 1957b. *Recursive Function Theory*. Amsterdam: North-Holland.

Gowers, T. 2002. *Mathematics: A Very Short Introduction.* Oxford: Oxford University Press.

Gowers, T. 2017a. *A Dialogue Concerning the Need for the Real Number System.* (At www.dpmms.cam.ac.uk/~wtg10/reals.html, accessed January 30, 2017.)

Gowers, T. 2017b. *What Is So Wrong with Thinking of Real Numbers as Infinite Decimals?* (At www.dpmms.cam.ac.uk/~wtg10/decimals.html, accessed January 30, 2017.)

Haller, R., & Puhl, K. (eds.). 2002. *Wittgenstein and the Future of Philosophy. A Reassessment after 50 Years.* Vienna: öbv&hpt.

Hardy, G. H. 1941. *A Course of Pure Mathematics* (8th ed.). Cambridge: Cambridge University Press. (10th ed. [1952, with index] republished as Hardy [2008].)

Hardy, G. H. 2008. *A Course of Pure Mathematics* (10th [1952, with index] ed.). Cambridge: Cambridge University Press.

Harrington, B., Shaw, D., & Beaney, M. (eds.). 2018. *Aspects after Wittgenstein: Seeing-As and Novelty.* New York: Routledge.

Hellman, G., & Cook, R. T. 2018. *Hilary Putnam on Logic and Mathematics* (vol. 9). Cham: Springer International Publishing AG Springer.

Herken, R. (ed.). 1988. *The Universal Turing Machine: A Half-Century Survey.* New York: Oxford University Press.

Hintikka, J. (ed.). 1995. *From Dedekind to Gödel: Essays on the Development of the Foundations of Mathematics.* Dordrecht: Kluwer Academic.

Hobson, E. W. 1927. *The Theory of Functions of a Real Variable and the Theory of Fourier's Series* (vol. I). Cambridge, UK: Cambridge University Press. (Revised and enlarged 3rd ed.; 1st ed., 1907, 2nd ed., 1921.)

Hrbacek, K., & Jech, T. J. 1999. *Introduction to Set Theory* (3rd revised and expanded ed.). Boca Raton: CRC / Taylor and Francis.

Irvine, A. (ed.). 2009. *Philosophy of Mathematics.* Amsterdam/Oxford: Elsevier.

Kennedy, J. 2017. *Turing, Gödel and the "Bright Abyss."* (In Floyd and Bokulich, 2017, pp. 63–92.)

Kennedy, J. 2020. *Gödel, Tarski and the Lure of Natural Language: Logical Entanglement, Formalism Freeness.* Cambridge, UK: Cambridge University Press.

Krebs, V., & Day, W. (eds.). 2010. *Seeing Wittgenstein Anew: New Essays on Aspect Seeing.* New York: Cambridge University Press.

Kripke, S. 1982. *Wittgenstein on Rules and Private Language: An Elementary Exposition.* Cambridge, MA: Harvard University Press.

Kuusela, O., & McGinn, M. (eds.). 2012. *The Oxford Handbook of Wittgenstein*. Oxford University Press.

Lillehammer, H., Mellor, D. H., & Mellor, D. H. 2005. *Ramsey's Legacy*. Oxford: Oxford University Press.

Maddy, P. 1997. *Naturalism in Mathematics*. Oxford, New York: Clarendon Press; Oxford University Press.

Maddy, P. 2005. *Three Forms of Naturalism*. (In Shapiro, 2005, pp. 437–459.)

Maddy, P. 2007. *Second Philosophy: A Naturalistic Method*. New York: Oxford University Press.

Maddy, P. 2011. *Defending the Axioms: On the Philosophical Foundations of Set Theory*. New York: Oxford University Press.

Maddy, P. 2014. *The Logical Must: Wittgenstein on Logic*. New York: Oxford University Press.

Makovec, D., & Shapiro, S. 2019. *Friedrich Waismann: The Open Texture of Analytic Philosophy*. Cham: Springer International Publishing AG Palgrave Macmillan.

Mancosu, P. 2008. *The Philosophy of Mathematical Practice*. New York: Oxford University Press.

Marion, M. 1998. *Wittgenstein, Finitism and the Foundations of Mathematics*. New York: Oxford, Clarendon Press.

Marion, M. 2003. Wittgenstein and Brouwer. *Synthese, 137*, 103–127.

Marion, M. 2007. Interpreting Arithmetic: Russell on Applicability and Wittgenstein on Surveyability. *Travaux de logique, 18*, 167–184.

Marion, M. 2011. *Wittgenstein on the Surveyability of Proofs*. (In Kuusela and McGinn, 2012, pp. 138–161.)

Marion, M., & Okada, M. 2018. *Wittgenstein, Goodstein and the Origin of the Uniqueness Rule*. (In Stern, 2018, pp. 253–271.)

Martin, C. (ed.). 2018. *Language, Form(s) of Life, and Logic: Investigations after Wittgenstein*. Berlin: de Gruyter.

Menger, K. 1994. *Reminiscences of the Vienna Circle and the Mathematical Colloquium* (Vol. 20). Dordrecht: Kluwer Academic/Springer Science+Business Media.

Monk, R. 1990. *Ludwig Wittgenstein: The Duty of Genius*. New York/London: Free Press/Jonathan Cape.

Moyal-Sharrock, D. 2015. Wittgenstein on Forms of Life, Patterns of Life, and Ways of Living. *Nordic Wittgenstein Review, Special Issue on Forms of Life*, 21–42.

Mühlhölzer, F. 2002. *Wittgenstein and Surprises in Mathematics*. (In Haller and Puhl, 2002, pp. 306–315.)

Mühlhölzer, F. 2006. "A Mathematical Proof Must Be Surveyable": What Wittgenstein Meant by This and What It Implies. *Grazer Philosophische Studien, 71*, 57–86.

Mühlhölzer, F. 2010. *Braucht die Mathematik eine Grundlegung? Ein Kommentar des Teils III von Wittgensteins BEMERKUNGEN ÜBER DIE GRUNDLAGEN DER MATHEMATIK.* Frankfurt am Main: Vittorio Klostermann. (English translation to appear in the Anthem Studies in Wittgenstein Series.)

Mühlhölzer, F. 2012. *Wittgenstein and Metamathematics.* (In Stekeler-Weithofer, 2012, pp. 103–128.)

Mühlhölzer, F. 2014. How Arithmetic Is about Numbers: A Wittgensteinian Perspective. *Grazer Philosophische Studien, 89*, 39–59.

Narboux, J.-P. 2014. Showing, the Medium Voice, and the Unity of the *Tractatus. Philosophical Topics, 42*(2).

Olszewski, A., Wolenski, J., & Janusz, R., (eds.). 2006. *Church's Thesis After 70 Years.* Frankfurt/Paris/Ebikon/Lancaster/New Brunswick: Ontos Verlag.

Paris, J., & Harrington, L. 1977. *A Mathematical Incompleteness in Peano Arithmetic.* (In Barwise, 1977, pp. 1133–1142.)

Parsons, C. 1983. Frege's Theory of Number. In *Mathematics in philosophy* (p. 150–175). Ithaca, NY: Cornell University Press.

Poincaré, H. 1904. *La Science et l'Hypothèse.* Flammarion.

Potter, M. 2000. *Reason's Nearest Kin: Philosophies of Arithmetic from Kant to Carnap.* New York: Oxford University Press.

Potter, M. 2005. *Ramsey's Transcendental Argument.* (In Lillehammer, Mellor, and Mellor, 2005, pp. 71–82.)

Putnam, H. 1962. It Ain't Necessarily So. *The Journal of Philosophy, LIX*, 658–671.

Putnam, H. 1990. *Realism with a Human Face.* Harvard University Press.

Putnam, H. 1992. *Renewing Philosophy.* Harvard University Press.

Putnam, H. 1999. *The Threefold Cord: Mind, Body and World* (vol. 112). Columbia University Press.

Putnam, H. 2004. *Ethics without Ontology.* Cambridge, MA: Harvard University Press.

Putnam, H. 2012. *Philosophy in an Age of Science: Physics, Mathematics, and Skepticism.* Cambridge, MA: Harvard University Press. (Edited by Mario De Caro and David Macarthur.)

Putnam, H. 2016. *Naturalism, Realism, and Normativity.* Cambridge, MA: Harvard University Press. (Edited by Mario De Caro.)

Quinon, P. 2020. Implicit and explicit examples of the phenomenon of deviant encodings. *Studies in Logic, Grammar and Rhetoric, 63*(76), 53–67.

Ramsey, F. P. 1923. Review of *Tractatus Logico-Philosophicus* by Ludwig Wittgenstein. *Mind, 32*(128), 465–478.

Ramsey, F. P. 1926. The Foundations of Mathematics. *Proceedings of the London Mathematical Society, Series 2, 25*(1), 338–384.

Ramsey, F. P. 1930. On a Problem of Formal Logic. *Proceedings of the London Mathematical Society, Series 2, 2–3*(1), 264–286.

Reck, E. H. (ed.). 2013. *Logic, Philosophy of Mathematics, and Their History: Essays in Honor of W. W. Tait* (Vol. 36). Milton Keynes, UK: College Publications.

Ricketts, T. 2014. Analysis, Independence, Simplicity and the General Sentence-Form. *Philosophical Topics, 42*(2), 263–288.

Rodych, V. 2002. Wittgenstein on Gödel: The Newly Published Remarks. *Erkenntnis, 56*(3), 379–397.

Rogers, B., & Wehmeier, K. 2012. Tractarian First-Order Logic: Identity and the N-Operator. *Review of Symbolic Logic, 5*(4), 538–573.

Russell, B. 1903. *The Principles of Mathematics.* Cambridge/New York: Cambridge University Press/Norton. (1st ed., 1903; 2nd ed., 1938.)

Russell, B. 1914. *Our Knowledge of the External World as a Field for Scientific Method in Philosophy.* Chicago: Open Court Publishing Co.

Russell, B. 1920. *Introduction to Mathematical Philosophy* (2nd ed.). London/New York: Allen and Unwin/Macmillan. (1st ed., 1919.)

Russell, B. 1936. The Limits of Empiricism. *Proceedings of the Aristotelian Society, 36*, 131–150.

Shanker, S. 1987. *Wittgenstein and the Turning-Point in the Philosophy of Mathematics.* Albany, NY: State University of New York Press.

Shapiro, S. 1982. Acceptable Notation. *Notre Dame Journal of Formal Logic, 23*(1), 14–20.

Shapiro, S. (ed.). 2005. *The Oxford Handbook to the Philosophy of Logic and Mathematics.* New York: Oxford University Press.

Shapiro, S. 2008. Identity, Indiscernibiity, and *ante rem* Structuralism: the Tale of i and $-i$. *Philosophia Mathematica, 16*, 285–309.

Shapiro, S. 2012. An "i" for an "i": Singular Terms, Uniqueness and Reference. *The Review of Symbolic Logic, 5*, 380–415.

Shapiro, S. 2018. Changing the Subject: Quine, Putnam and Waismann on Meaning-Change, Logic, and Analyticity. (In Hellman and Cook, 2018, pp. 115–126.)

Sheffer, H. M. 1913. A set of five independent postulates for Boolean Algebras, with Application to Logical Constants. *Transactions of the American Mathematical Society, 14*(4), 481–488.

Shieh, S. 2019. *Necessity Lost. Modality and Logic in Early Analytic Philosophy. Volume I* (1st ed.). Oxford: Oxford University Press.

Sieg, W. 2009. On Computability. (In Irvine, 2009, pp. 535–630.)

Skolem, T. 1923. *The foundations of elementary arithmetic established by means of the recursive mode of thought, without the use of apparent variables ranging over infinite domains.* (In van Heijenoort, 1967, pp. 302–333)

Sluga, H., & Stern, D. (eds.). 2018. *The Cambridge Companion to Wittgenstein* (2nd rev. ed.). New York: Cambridge University Press.

Snyder, E., & Shapiro, S. 2019. *Frege on the Real Numbers.* (In Ebert and Rossberg, 2019, pp. 343–383.)

Steiner, M. 1975. *Mathematical Knowledge.* Cornell University Press.

Stekeler-Weithofer, P. (ed.). 2012. *Wittgenstein: Zu Philosophie und Wissenschaft.* Hamburg: Verlag Felix Meiner.

Stern, D. G. (ed.). 2018. *Wittgenstein in the 1930s: Between the Tractatus and the Investigations.* Cambridge: Cambridge University Press.

Sullivan, P. 1995. Wittgenstein on The Foundations of Mathematics, June 1927. *Theoria, 61*, 105–142.

Sundholm, G. 1992. The General Form of the Operation in Wittgenstein's Tractatus. *Grazer Philosophische Studien, 42*, 57–76.

Tao, T. 2006. *Analysis I.* Gurugram, Hindustan India: Hindustan Book Agency.

Travis, C. 2006. *Thought's Footing: A Theme in Wittgenstein's Philosophical Investigations.* Oxford/New York: Oxford University Press.

Turing, A. 1936. On computable numbers, with an application to the *Entscheidungsproblem. Proceedings of the London Mathematical Society, Series 2, 1936/7*(42), 230–265. Corrections in Turing, 1937.

Turing, A. 1937. On computable numbers, with an application to the *Entscheidungsproblem*: A Correction. *Proceedings of the London Mathematical Society, Series 2, 1937*(43), 544–546.

Turing, A. 1942–44. *The Reform of Mathematical Notation and Phraseology [1944-45].* (Unpublished manuscript. In Turing, 2013, pp. 245–249; online at www.turingarchive.org/search/, King's College Online Archive.)

Turing, A. 1948. *Intelligent Machinery.* (Unpublished report for the National Physical Laboratory. In Turing, 2013, pp. 501–516.)

Turing, A. 2013. *Alan Turing – His Work and Impact.* Amsterdam/Burlington, MA: Elsevier. (Edited by S. Barry Cooper and Jan van Leeuwen; collected papers of Turing with commentary by experts.)

van Heijenoort, J. (ed.). 1967. *From Frege to Gödel: A Source Book in Mathematical Logic, 1879–1931.* Cambridge, MA: Harvard University Press. (Reprinted 2002)

Van Heijenoort, J. 1967. Logic as Calculus and Logic as Language. *Synthese,* *17*, 324–330.

Waismann, F. 1936a. *Einführung in das mathematische Denken.* Vienna: Springer. (English translation by Theodore J. Benac, *Introduction to Mathematical Thinking: The Formation of Concepts in Modern Mathematics.* New York: Harper, 1959; Dover reprint 2003, references to this edition.)

Waismann, F. 1936b. Über den begriff der identität. *Erkenntnis, 6*(1), 56–64.

Waismann, F. 1976/1997. *The Principles of Linguistic Philosophy* (2nd ed.). London: Macmillan.

Waismann, F. 1982. *Lectures on the Philosophy of Mathematics.* Amsterdam: Rodopi.

Weiss, M. 2017. Logic in the *Tractatus. Review of Symbolic Logic, 10*(1), 1–50.

Weyl, H. 1918. *Das Kontinuum: Kritische Untersuchungen über die Grundlagen der Analysis.* Leipzig: Verlag von Veit und Comp. (English translation by S. Pollard and T. Bole, *The Continuum*, Dover Publications, New York: 1994. References are to this translation.)

Whitehead, A. N., & Russell, B. 1910. *Principia Mathematica.* Cambridge: University Press. (3 vols., 2nd ed., 1927.)

Wittgenstein, L. 1974. *Philosophical Grammar.* Oxford: Basil Blackwell. (Edited by Rush Rhees, English translation by Anthony Kenny.)

Wittgenstein, L., & Waismann, F. 2003. *The Voices of Wittgenstein, the Vienna Circle: Original German Texts and English Translations.* London: Routledge. (Edited by Gordon P. Baker.)

Wright, C. 1980. *Wittgenstein on the Foundations of Mathematics.* Harvard University Press.

Wright, C. 2001. *Rails to Infinity: Essays on Themes from Wittgenstein's Philosophical Investigations.* Cambridge, MA: Harvard University Press.

Acknowledgments

Mauro Engelmann and Sanford Shieh insightfully commented on early versions of this manuscript, suggesting many important improvements, with Sanford framing for me the current debate over second-order generalization in TLP. Zeynep Soysal and Juliette Kennedy inspired me with their writings in philosophy of mathematics, giving me crucial feedback on the material in §4.5, Juliette especially on the final manuscript. Felix Mühlhölzer provided me thorough comments on the final manuscript and years of conversation, friendship, and breakthrough writing about the Later Wittgenstein's views of mathematics.

The Boston University Center for the Humanities, under the direction of Susan Mizruchi, generously funded my time to write this Element.

The Philosophy of Mathematics

Penelope Rush

University of Tasmania

From the time Penny Rush completed her thesis in the philosophy of mathematics (2005), she has worked continuously on themes around the realism/anti-realism divide and the nature of mathematics. Her edited collection *The Metaphysics of Logic* (Cambridge University Press, 2014), and forthcoming essay 'Metaphysical Optimism' (*Philosophy Supplement*), highlight a particular interest in the idea of reality itself and curiosity and respect as important philosophical methodologies.

Stewart Shapiro

The Ohio State University

Stewart Shapiro is the O'Donnell Professor of Philosophy at The Ohio State University, a Distinguished Visiting Professor at the University of Connecticut, and a Professorial Fellow at the University of Oslo. His major works include *Foundations without Foundationalism* (1991), *Philosophy of Mathematics: Structure and Ontology* (1997), *Vagueness in Context* (2006), and *Varieties of Logic* (2014). He has taught courses in logic, philosophy of mathematics, metaphysics, epistemology, philosophy of religion, Jewish philosophy, social and political philosophy, and medical ethics.

About the Series

This Cambridge Elements series provides an extensive overview of the philosophy of mathematics in its many and varied forms. Distinguished authors will provide an up-to-date summary of the results of current research in their fields and give their own take on what they believe are the most significant debates influencing research, drawing original conclusions.

Cambridge Elements ☰

The Philosophy of Mathematics

Printed in the United States
by Baker & Taylor Publisher Services